T0155515

SpringerBriefs in Materials

The SpringerBriefs Series in Materials presents highly relevant, concise monographs on a wide range of topics covering fundamental advances and new applications in the field. Areas of interest include topical information on innovative, structural and functional materials and composites as well as fundamental principles, physical properties, materials theory and design. SpringerBriefs present succinct summaries of cutting-edge research and practical applications across a wide spectrum of fields. Featuring compact volumes of 50 to 125 pages, the series covers a range of content from professional to academic. Typical topics might include:

- A timely report of state-of-the-art analytical techniques
- A bridge between new research results, as published in journal articles, and a contextual literature review
- A snapshot of a hot or emerging topic
- An in-depth case study or clinical example
- A presentation of core concepts that students must understand in order to make independent contributions

Briefs are characterized by fast, global electronic dissemination, standard publishing contracts, standardized manuscript preparation and formatting guidelines, and expedited production schedules.

More information about this series at http://www.springer.com/series/10111

Geon Dae Moon

Anisotropic Metal Chalcogenide Nanomaterials

Synthesis, Assembly, and Applications

Geon Dae Moon
Dongnam Regional Division
Korea Institute of Industrial Technology
Busan, Korea (Republic of)

ISSN 2192-1091 ISSN 2192-1105 (electronic)
SpringerBriefs in Materials
ISBN 978-3-030-03942-4 ISBN 978-3-030-03943-1 (eBook)
https://doi.org/10.1007/978-3-030-03943-1

Library of Congress Control Number: 2018961199

This Springer imprint is published by the registered company Springer Nature Switzerland AG
The registered company address is: Gewerbestrasse 11, 6330 Cham, Switzerland

Dedicated to my life partner, Y. Kim
and daughter, J. Moon.

Preface

This book deals with 1D and 2D anisotropic metal chalcogenide (MC) nanocrystals with a focus on the solution-based synthesis and their practical and/or future applications. Different growth mechanisms to control the shapes of the MCs will be reviewed: spontaneous growth due to its intrinsic anisotropic crystal structure, shape-guided anisotropic growth by changing surface energies or utilizing organic templates, axially or laterally oriented attachment of small nanocrystal building blocks to form nanowires or nanosheets, and, finally, chemical transformation from existing nanostructures into new species. We discuss current understanding of the thermodynamic and kinetic aspects associated with the mechanisms of forming these anisotropic MC nanostructures. Quite recently, ultrathin films of MCs are on the center of the MC research due to its atomically thin quantum nature with their availability of large-area fabrication. Representative examples wait for the readers to bridge the gap between design/synthesis and applications by using anisotropic MC nanomaterials that are expected to be meaningful practically in the near future. The applications include electrodes for energy storage and conversion, memory device, photodetectors, electrocatalyst, topological insulator, localized surface plasmon resonance, and superconductor. This book ends with discussions on the challenges and perspectives to be investigated thoroughly in the design of anisotropic MC nanomaterials, geared toward application-targeted synthesis, scaled-up production, and environmentally friendly process.

Busan, Korea Geon Dae Moon

About This Book

This brief deals with recent advances of 1D and 2D anisotropic metal chalcogenide (MC) nanostructures and addresses their synthetic approach, assembly, properties, and practical area of applications in electronics, energy storage/conversion, sensor, and catalyst. The primary focus is on the current understanding of the thermodynamic and kinetic aspects associated with the mechanisms of forming these anisotropic MC nanostructures. Control over the shape and size of nanocrystals has always been desired and essential to tune their chemical or physical properties for target applications. The size of nanocrystals is proven to exhibit conspicuously peculiar characteristics compared to their bulk counterparts. Similarly, anisotropic nanocrystals have unique properties that are direction-dependent and the ability to confine the motion of electrons, holes, excitons, phonons, and plasmons in specific and controlled directions. Not only an overview of background and principle, but also representative applications of anisotropic metal chalcogenide nanomaterials are provides, which is expected to meaningful in industrial sectors in the near future.

Contents

About the Author

Dr. Geon Dae Moon is now working as a senior researcher at Korea Institute of Industrial Technology (KITECH) in Busan, Korea. He has obtained his Ph.D. of Materials Science and Engineering from Yonsei University, Seoul, Korea. Then, he worked as a postdoctoral researcher in chemistry, University of California, Riverside, and as a researcher associate in the Sustainable Energy Technology Department, Brookhaven National Laboratory. His research interest includes the synthesis of nanostructures based on chemistry and assembly of nanomaterials toward electronic, energy, and environment applications.

Chapter 1
Introduction

Abstract Nanoscience and nanotechnology have been putting the size and shape-dependent characteristics of matters on the horizon. Due to this reason, huge efforts have been poured into finding parameters affecting the intrinsic properties of materials at nanometer scale. Other than the size effect, shapes of nanomaterials produce a class of materials in which their properties are variable with direction. The anisotropy in chemical and physical features created vigorous attention in a way to synthesize 1D and 2D nanomaterials. Especially, metal chalcogenide (MC), constituting a series of compound materials with its vast library on the periodic table, has been studied and demonstrated to exploit in numerous applications.

Having the ability to exert a control over the shape of nanocrystals has always been desired and essential to tune their chemical or physical properties for target applications [1]. Anisotropic nanocrystals have unique properties that are direction-dependent and have the ability to confine the motion of electrons, holes, excitons, phonons, and plasmons in specific and controlled directions [2]. Such uniqueness have been employed in electronics (conducting platforms, transistors, electromechanical devices) [3–6], energy conversion and storage devices (Li ion batteries, solar cells, thermoelectrics, piezoelectrics) [7–13], optical devices (lasers, OLED, photodetectors) [14–17], and electrochemical devices (catalysis, gas sensors) [18, 19]. Anisotropic nanocrystals have been observed in many materials including carbons, silicon, metals, metal oxides, chalcogenides, carbides, nitrides, and their compounds [20, 21].

Low dimensional metal chalcogenide (MC) nanostructures have attracted considerable attention from chemistry, physics, and materials science since the discovery of carbon nanotubes and exfoliation of graphene [22, 23]. The active research efforts are inevitable due to the peculiar chemical and physical properties of anisotropic nanomaterials that change significantly from their bulk counterparts. Thus, there seems to have numerous opportunities to engineer unique properties by transforming shapes, size, crystallinity, and composition for never-achieved development of advanced functional devices in various applications. For example, two-dimensional MC nanocrystals are also subject to strong quantum

G. D. Moon, *Anisotropic Metal Chalcogenide Nanomaterials*,
SpringerBriefs in Materials, https://doi.org/10.1007/978-3-030-03943-1_1

confinement compared to their 3D materials such as significant changes of spontaneous monolayer rippling, the electronic band structure, giant spin-orbit splitting, and enhanced photoluminescence leading to different structural, electronic, and optical properties [24]. Specifically, single-layer MoS_2 is a direct band gap semiconductor with spin-orbit splitting of 150 meV in the valence band, while the bilayer is an indirect band gap semiconductor without spin-orbit splitting. Large surface-to-volume ratio of anisotropic MC nanomaterials enables the control over multi-component MC nanostructure synthesis through rational chemical reaction. Meanwhile, the bulk counterparts limit the chemical reactivity around the thin layers of the surfaces. With the chemical and physical properties, the potential applications of anisotropic MC nanostructures have paved its way towards electronics, energy conversion/storage, optics, and chemical reactivity [25–30]. For the realization of the above devices, individual MC nanocrystals should be incorporated with other components without losing the properties of discrete MCs. As a general processing technique, colloidal self-assembly has been, vigorously, studied by exploiting various forces including intermolecular force, electrostatic force, capillary force, and convective motion of solvents. Some interesting phenomena can occur during 1D and 2D nanostructures' assembly, opening up new opportunities for real applications. Heterostructured MC nanocrystals can also synthesized with dislocation-driven self-assembly or epitaxial growth on the template MC nanostructures [31].

Last, but not least, the large area thin film of MC nanocrystal (i.e., wafer-scale) is enabled by layer-by-layer assembly of 2D nanostructure, geared towards high performance semiconductor heterojunction films [33].

Although it was only recent that anisotropic metal chalcogenide (MC) nanocrystals have been investigated with much interest (with regards to their applications), remarkable advances have been made with anisotropic nanocrystals in diverse areas of modern technology [25]. These MC nanocrystals are also strong candidate materials for thermoelectric devices. Energy harvested from heat loss and cooling electronic devices by effective heat transfer have motivated the studies on the chalcogenide-based thermoelectric devices [34]. In photovoltaic and photodector devices, MC nanocrystals have shown to absorb sun light, hence they can play as exciton generators and charge transport materials [35–37]. The MC nanocrystals with a layered crystal structure and their composites with carbon materials are promising candidates that can intercalate Li^+ ions reversibly, which can be used as electrodes for batteries [38, 39]. In addition to these applications, new potential novel applications are recently being investigated for these MCs. As an alternative to noble metals, doped MC nanostructure has attracted interest as an efficient catalyst for the oxygen reduction reaction [40, 41] and as a patternable material whose localized surface plasmon resonance is tunable [42, 43]. The topological insulating characteristics [44] and superconductivity [45] of MC nanostructures are also considered new areas of research for this class of materials.

Anisotropic MC nanocrystals have predominantly been synthesized in solutions because the size and shape of the nanocrystals can be controlled precisely and the solution-processed printing is the desired inexpensive way to produce MC films

[46–49]. The well-known Gibbs-Wulff theorem determines the shape of nanocrystal in way that the total surface energy of the crystal faucets minimizes [50]. Chalcogens (S, Se, Te) and some MCs have the preferred directional growth, which means the existence of thermodynamically-stable crystal growth. Se and Te are well known to grow into one-dimensional nanostructures due to their strong bond along *c*-axis [51]. 2D MCs such as Bi_2Se_3 and Bi_2Te_3 can grow in a thermodynamic preference without any shape-guiding agents. Table 1.1 summarizes the thermodynamic preference in shape and crystal structure of MC nanocrystals [32]. It varies in the metal species and the stoichiometric ratio of the metal to chalcogen (M_xQ_y, M = metal, Q = S, Se, Te). These materials, according to each category, are listed and the tendency to form the anisotropic shape is noted as 'strong' or 'medium' in the table.

This book introduces the common solution-based synthesis of anisotropic MC nanocrystals and the recent advances in applications, with some guidance from first-principles simulations and computations. The synthetic pathways of 1D and 2D structures are dealt with in separate sections. We begin with the direct synthesis from organometal precursors and then review the chemical transformation of the premade nanocrystals into others. The application section mainly focuses on the

Table 1.1 Thermodynamic preferences of chalcogenide nanocrystals

	M (metal)	Thermodynamic preference	Crystal structure	Reason	MCs
Q (=S, Se, Te)		1D (strong)	Trigonal	Chain-like structure	Se, Te
M_2Q	IB	1D (medium)	Hexagonal	High surface E in the (0001)	Cu_2Q
		0D (strong)	Cubic	Isotropic surface E	$Cu_{2-x}Q$
MQ	IIB, IVA	0D (strong)	Cubic	Isotropic surface E	ZnQ, CdQ, PbQ
	IIB	1D (strong)	Hexagonal	High surface E in the (0001)	CdQ, ZnQ
	IVA, VIIIB	2D (strong)	Orthorhombic	Layered structure	GeQ, SnQ, FeQ
MQ_2	IB, IIB, VIIIB	3D (medium)	Cubic	Isotropic surface E	FeS_2, CoS_2, NiS_2, CuS_2
	IVB, VB, VIB	2D (strong)	Hexagonal	Layered structure	TiQ_2, ZrQ_2, NbQ_2, MoQ_2
MQ_3	IVB	1D (medium)	Monoclinic	Chain-like structure	TiS_3, ZrQ_3, HfQ_3
M_2Q_3	VA	2D (strong)	Hexagonal	Layered structure	Bi_2Q_3, Sb_2Q_3
		1D (strong)	Orthorhombic	Chain-like structure	Bi_2Q_3, Sb_2Q_3

Reproduced by permission of the Royal Society of Chemistry [32]

research areas which can be practically meaningful in the near future, and then introduces a few emerging topics in physics and chemistry. We conclude this brief with discussion on the challenges and future directions in the solution-based synthesis and their applications.

References

1. Burda C, Chen X, Narayanan R, El-Sayed MA (2005) Chemistry and properties of nanocrystals of different shapes. Chem Rev 105(4):1025–1102. https://doi.org/10.1021/cr030063a
2. Sajanlal PR, Sreeprasad TS, Samal AK, Pradeep T (2011) Anisotropic nanomaterials: structure, growth, assembly, and functions. Nano Rev 2(1):5883. https://doi.org/10.3402/nano.v2i0.5883
3. Zhu J, Shim BS, Di Prima M, Kotov NA (2011) Transparent conductors from carbon nanotubes LBL-assembled with polymer dopant with—electron transfer. J Am Chem Soc 133 (19):7450–7460. https://doi.org/10.1021/ja111687t
4. Azulai D, Belenkova T, Gilon H, Barkay Z, Markovich G (2009) Transparent metal nanowire thin films prepared in mesostructured templates. Nano Lett 9(12):4246–4249. https://doi.org/10.1021/nl902458j
5. Lee J-Y, Connor ST, Cui Y, Peumans P (2008) Solution-processed metal nanowire mesh transparent electrodes. Nano Lett 8(2):689–692. https://doi.org/10.1021/nl073296g
6. De S, Lyons PE, Sorel S, Doherty EM, King PJ, Blau WJ, Nirmalraj PN, Boland JJ, Scardaci V, Joimel J, Coleman JN (2009) Transparent, flexible, and highly conductive thin films based on polymer–nanotube composites. ACS Nano 3(3):714–720. https://doi.org/10.1021/nn800858w
7. Aricò AS, Bruce P, Scrosati B, Tarascon J-M, van Schalkwijk W (2005) Nanostructured materials for advanced energy conversion and storage devices. Nat Mater 4:366. https://doi.org/10.1038/nmat1368
8. Li H, Wang Z, Chen L, Huang X (2009) Research on advanced materials for Li-ion Batteries. Adv Mater 21(45):4593–4607. https://doi.org/10.1002/adma.200901710
9. Wang X, Zhi L, Müllen K (2008) Transparent, conductive graphene electrodes for dye-sensitized solar cells. Nano Lett 8(1):323–327. https://doi.org/10.1021/nl072838r
10. Hochbaum AI, Yang P (2010) Semiconductor nanowires for energy conversion. Chem Rev 110(1):527–546. https://doi.org/10.1021/cr900075v
11. Harman TC, Taylor PJ, Walsh MP, LaForge BE (2002) Quantum dot superlattice thermoelectric materials and devices. Science 297(5590):2229–2232. https://doi.org/10.1126/science.1072886
12. Hsu KF, Loo S, Guo F, Chen W, Dyck JS, Uher C, Hogan T, Polychroniadis EK, Kanatzidis MG (2004) Cubic $AgPbmSbTe_{2+}$ m: bulk thermoelectric materials with high figure of merit. Science 303(5659):818–821. https://doi.org/10.1126/science.1092963
13. Wang ZL (2008) Towards self-powered nanosystems: from nanogenerators to nanopiezotronics. Adv Func Mater 18(22):3553–3567. https://doi.org/10.1002/adfm.200800541
14. Cho C-H, Aspetti CO, Turk ME, Kikkawa JM, Nam S-W, Agarwal R (2011) Tailoring hot-exciton emission and lifetimes in semiconducting nanowires via whispering-gallery nanocavity plasmons. Nat Mater 10:669. https://doi.org/10.1038/nmat3067
15. Schuller JA, Barnard ES, Cai W, Jun YC, White JS, Brongersma ML (2010) Plasmonics for extreme light concentration and manipulation. Nat Mater 9:193. https://doi.org/10.1038/nmat2630

16. Ou ECW, Hu L, Raymond GCR, Soo OK, Pan J, Zheng Z, Park Y, Hecht D, Irvin G, Drzaic P, Gruner G (2009) Surface-modified nanotube anodes for high performance organic light-emitting diode. ACS Nano 3(8):2258–2264. https://doi.org/10.1021/nn900406n
17. McDonald SA, Konstantatos G, Zhang S, Cyr PW, Klem EJD, Levina L, Sargent EH (2005) Solution-processed PbS quantum dot infrared photodetectors and photovoltaics. Nat Mater 4:138. https://doi.org/10.1038/nmat1299
18. Lim B, Jiang M, Camargo PHC, Cho EC, Tao J, Lu X, Zhu Y, Xia Y (2009) Pd-Pt bimetallic nanodendrites with high activity for oxygen reduction. Science 324(5932):1302–1305. https://doi.org/10.1126/science.1170377
19. Kauffman DR, Sorescu DC, Schofield DP, Allen BL, Jordan KD, Star A (2010) Understanding the sensor response of metal-decorated carbon nanotubes. Nano Lett 10 (3):958–963. https://doi.org/10.1021/nl903888c
20. Rao CNR, Deepak FL, Gundiah G, Govindaraj A (2003) Inorganic nanowires. Prog Solid State Chem 31(1):5–147. https://doi.org/10.1016/j.progsolidstchem.2003.08.001
21. Wang X, Li Y (2006) Solution-based synthetic strategies for 1-D nanostructures. Inorg Chem 45(19):7522–7534. https://doi.org/10.1021/ic051885o
22. Iijima S (1991) Helical microtubules of graphitic carbon. Nature 354:56. https://doi.org/10.1038/354056a0
23. Novoselov KS, Geim AK, Morozov SV, Jiang D, Zhang Y, Dubonos SV, Grigorieva IV, Firsov AA (2004) Electric field effect in atomically thin carbon films. Science 306 (5696):666–669. https://doi.org/10.1126/science.1102896
24. Heine T (2015) Transition metal chalcogenides: ultrathin inorganic materials with tunable electronic properties. Acc Chem Res 48(1):65–72. https://doi.org/10.1021/ar500277z
25. Shi W, Hughes RW, Denholme SJ, Gregory DH (2010) Synthesis design strategies to anisotropic chalcogenide nanostructures. CrystEngComm 12(3):641–659. https://doi.org/10.1039/B918794B
26. Gong C, Zhang Y, Chen W, Chu J, Lei T, Pu J, Dai L, Wu C, Cheng Y, Zhai T, Li L, Xiong J (2017) Electronic and optoelectronic applications based on 2D novel anisotropic transition metal dichalcogenides. Adv Sci 4(12):1700231. https://doi.org/10.1002/advs.201700231
27. Bi W, Zhou M, Ma Z, Zhang H, Yu J, Xie Y (2012) $CuInSe_2$ ultrathin nanoplatelets: novel self-sacrificial template-directed synthesis and application for flexible photodetectors. Chem Commun 48(73):9162–9164. https://doi.org/10.1039/C2CC34727J
28. Min Y, Roh JW, Yang H, Park M, Kim SI, Hwang S, Lee SM, Lee KH, Jeong U (2013) Surfactant-free scalable synthesis of Bi_2Te_3 and Bi_2Se_3 nanoflakes and enhanced thermoelectric properties of their nanocomposites. Adv Mater 25(10):1425–1429. https://doi.org/10.1002/adma.201203764
29. Seo JW, Jang JT, Park SW, Kim C, Park B, Cheon J (2008) Two-dimensional SnS_2 nanoplates with extraordinary high discharge capacity for lithium ion batteries. Adv Mater 20 (22):4269–4273. https://doi.org/10.1002/adma.200703122
30. Gao MR, Jiang J, Yu SH (2012) Solution-based synthesis and design of late transition metal chalcogenide materials for oxygen reduction reaction (ORR). Small 8(1):13–27. https://doi.org/10.1002/smll.201101573
31. Dun C, Hewitt CA, Li Q, Xu J, Schall DC, Lee H, Jiang Q, Carroll DL (2017) 2D chalcogenide nanoplate assemblies for thermoelectric applications. Adv Mater 29 (21):1700070. https://doi.org/10.1002/adma.201700070
32. Min Y, Moon GD, Kim C-E, Lee J-H, Yang H, Soon A, Jeong U (2014) Solution-based synthesis of anisotropic metal chalcogenide nanocrystals and their applications. J Mater Chem C 2(31):6222–6248. https://doi.org/10.1039/C4TC00586D
33. Kang K, Lee K-H, Han Y, Gao H, Xie S, Muller DA, Park J (2017) Layer-by-layer assembly of two-dimensional materials into wafer-scale heterostructures. Nature 550:229. https://doi.org/10.1038/nature23905
34. Bell LE (2008) Cooling, heating, generating power, and recovering waste heat with thermoelectric systems. Science 321(5895):1457–1461. https://doi.org/10.1126/science.1158899

35. Talapin DV, Lee J-S, Kovalenko MV, Shevchenko EV (2010) Prospects of colloidal nanocrystals for electronic and optoelectronic applications. Chem Rev 110(1):389–458. https://doi.org/10.1021/cr900137k
36. Zhong H, Lo SS, Mirkovic T, Li Y, Ding Y, Li Y, Scholes GD (2010) noninjection gram-scale synthesis of monodisperse pyramidal $CuInS_2$ nanocrystals and their size-dependent properties. ACS Nano 4(9):5253–5262. https://doi.org/10.1021/nn1015538
37. He M, Qiu F, Lin Z (2013) Toward high-performance organic-inorganic hybrid solar cells: bringing conjugated polymers and inorganic nanocrystals in close contact. J Phys Chem Lett 4 (11):1788–1796. https://doi.org/10.1021/jz400381x
38. Wang X, Chen D, Yang Z, Zhang X, Wang C, Chen J, Zhang X, Xue M (2016) Novel metal chalcogenide SnSSe as a high-capacity anode for sodium-ion batteries. Adv Mater 28 (39):8645–8650. https://doi.org/10.1002/adma.201603219
39. Hwang H, Kim H, Cho J (2011) MoS_2 nanoplates consisting of disordered graphene-like layers for high rate lithium battery anode materials. Nano Lett 11(11):4826–4830. https://doi.org/10.1021/nl202675f
40. Chen P, Xiao T-Y, Li H-H, Yang J-J, Wang Z, Yao H-B, Yu S-H (2012) Nitrogen-doped Graphene/ZnSe manocomposites: hydrothermal synthesis and their enhanced electrochemical and photocatalytic activities. ACS Nano 6(1):712–719. https://doi.org/10.1021/nn204191x
41. Liu S, Zhang Z, Bao J, Lan Y, Tu W, Han M, Dai Z (2013) Controllable synthesis of tetragonal and cubic phase Cu_2Se nanowires assembled by small nanocubes and their electrocatalytic performance for oxygen reduction reaction. J Phys Chem C 117(29):15164–15173. https://doi.org/10.1021/jp4044122
42. Luther JM, Jain PK, Ewers T, Alivisatos AP (2011) Localized surface plasmon resonances arising from free carriers in doped quantum dots. Nat Mater 10:361. https://doi.org/10.1038/nmat3004
43. Hsu S-W, On K, Tao AR (2011) Localized surface plasmon resonances of anisotropic semiconductor nanocrystals. J Am Chem Soc 133(47):19072–19075. https://doi.org/10.1021/ja2089876
44. Xiu F, He L, Wang Y, Cheng L, Chang L-T, Lang M, Huang G, Kou X, Zhou Y, Jiang X, Chen Z, Zou J, Shailos A, Wang KL (2011) Manipulating surface states in topological insulator nanoribbons. Nat Nanotechnol 6:216. https://doi.org/10.1038/nnano.2011.19
45. Ren ZA, Zhao ZX (2009) Research and prospects of iron-based superconductors. Adv Mater 21(45):4584–4592. https://doi.org/10.1002/adma.200901049
46. Jiang C, Lee J-S, Talapin DV (2012) Soluble precursors for $CuInSe_2$, $CuIn1{-}xGaxSe_2$, and $Cu_2ZnSn(S, Se)_4$ based on colloidal nanocrystals and molecular metal chalcogenide surface ligands. J Am Chem Soc 134(11):5010–5013. https://doi.org/10.1021/ja2105812
47. Giri A, Park G, Yang H, Pal M, Kwak J, Jeong U (2018) Synthesis of 2D metal chalcogenide thin films through the process involving solution phase deposition. Adv Mater 0 (0):1707577. https://doi.org/10.1002/adma.201707577
48. Steinhagen C, Panthani MG, Akhavan V, Goodfellow B, Koo B, Korgel BA (2009) Synthesis of Cu_2ZnSnS_4 nanocrystals for use in low-cost photovoltaics. J Am Chem Soc 131 (35):12554–12555. https://doi.org/10.1021/ja905922j
49. Guo Q, Hillhouse HW, Agrawal R (2009) Synthesis of Cu_2ZnSnS_4 nanocrystal ink and its use for solar cells. J Am Chem Soc 131(33):11672–11673. https://doi.org/10.1021/ja904981r
50. Venables JA (2003) Introduction to surface and thin film processes. Cambridge University Press, Cambridge
51. Gates B, Mayers B, Cattle B, Xia Y (2002) Synthesis and characterization of uniform nanowires of trigonal selenium. Adv Func Mater 12(3):219–227. https://doi.org/10.1002/1616-3028(200203)12:3%3c219:AID-ADFM219%3e3.0.CO;2-U

Chapter 2
Synthesis and Assembly

Abstract The synthetic protocol for anisotropic MC nanocrystals has been developed based on accumulation of a largely empirical recipe, followed by inductive conclusion. Afterwards, the initial parameters are optimized to obtain the targeted nanostructures. The systematic synthetic effort is repeated and devised to inter-relate each parameter with a rational design of novel and complex MC nanostructures. Normally, the shape of nanocrystals obtained through thermodynamically-controlled growth reflects the inherent symmetry of the crystal structure, hence 3D nanocrystals are readily obtained. Synthesis of nanocrystals that do not have any preferential growth direction requires purposeful shape-guiding. Reduction of the surface energy of a certain facet is a powerful way to obtain anisotropic MCs, which can be achieved by selectively attaching organic surfactants or using the organic templates as the shape-determining reactor. The MC nanocrystals can then be merged to form 1D or 2D nanostructures. Such oriented attachment can be tailored by engineering the interaction between the nanocrystals. Chemical transformation of pre-existing anisotropic nanocrystals into others has recently received a lot of interest because it allows preparation of nanocrystals that are chemically different but have the same shape and dimensions.

The synthesis is normally based on the inherent preference in a certain growth direction or the surfactant-driven shape guiding. Chalcogens (S, Se, Te) and certain MCs are representative material species having the preferred directional growth. The shape of nanocrystals is known to be dominated by the total minimum surface energy of the crystal facets; this is known as the Gibbs-Wulff theorem Venables [1]. In many chalcogenides, the minimized total surface energies are found in the 1D shapes because of the asymmetric bond strength. Se and Te are well known to grow into nanowires or nanotubes [2]. The layer-structured chalcogenides grow into 2D nanoplates or nanosheets, as frequently observed in Bi_2Se_3, Bi_2Te_3, MoS_2, and SnS_2, etc. The assembly technique has been developed in parallel with the synthetic advance to achieve an ordered arrangement of MC nanocrystals as 2D structures or heterostructures with an aim for realization of devices. In this chapter, various

G. D. Moon, *Anisotropic Metal Chalcogenide Nanomaterials*,
SpringerBriefs in Materials, https://doi.org/10.1007/978-3-030-03943-1_2

synthetic protocols of 1D and 2D anisotropic MCs will be introduced, followed by techniques for assembly of those MC nanocrystals for real applications.

2.1 Synthetic Strategies for 1D Metal Chalcogenides

One-dimensional (1D) nanostructures include nanowire, nanorod, nanotube, nanobelt, nanoneedle, nanoribbon, nanofiber, whisker, etc. For the formation of anisotropic nanostructure, the crystal grows along a certain orientation faster than the others. Especially, nanowires with uniform diameter can be obtained when crystal grows along one direction with no or suppressed growth along the other directions. Periodic bond chain (PBC) theory explains the thermodynamically equilibrium crystal based on different growth rate (surfaces with high surface energy grow faster and disappear finally) [56]. Thus, only surfaces with the lowest total surface energy will survive. Thus, the formation of 1D nanostructures depending on the different growth rate is limited to some materials with special crystal structures. Furthermore, a low supersaturation is essential for anisotropic growth, otherwise, secondary or homogeneous nucleation can occur at a high supersaturation. Nevertheless, numerous 1D MC have been synthesized by modulating the surface energy with surfactant or attachment. In addition, intrinsically-grown 1D chalcogens or metal chalcogenides can be exploited as chemical or physical template to produce other MCs through chemical transformation. The representative 1D MC nanocrystals are tabulated in Table 2.1. The reported synthetic protocols can be classified into four different pathways:

1. Intrinsic growth
2. Shape-guiding agent growth
3. Oriented attachment
4. Chemical transformation.

The following sub-chapters will deal with the above synthetic routes for 1D MC nanocrystals mainly with recent developments.

2.1.1 Intrinsic Growth

Some solid materials inherently grow into anisotropic nanostructures. Chalcogens (e.g., Se, Te) are well-known examples. Se or Te atoms are covalently bonded in a helical chain along the *c*-axis, while the helical chains are hexagonally closely packed through van der Waals interaction. The stronger atomic bond strength in the *c*-axis results in rapid growth in the same direction. Figure 2.1a shows the crystal structure of Te with the lattice spacings calculated by density-functional theory (DFT) with Grimme's van der Waals correction. [57] We performed the DFT

Table 2.1 1D metal chalcogenide nanocrystals reported to date

Mechanism	Product		Morphology
Intrinsic growth	Se, Te, Se_xTe_y		NW, [2, 4–14] NT, [8, 15] NR, [7, 16, 17] NN, [18] NRB, [19] NB [15]
	Bi_2S_3		NW, NR [20]
	Sb_2Q_3 (Q = S, Se)		NW, [21] NRB [22]
Shape-guiding agent growth	Cu_2Q (Q = S, Te), $Cu_{2-x}Q$		NW [23]
	PbQ (Q = Se, Te)		NW, [24, 25] NR [24, 25]
	Sb_2Te_3		NB [26]
Axial oriented attachment	PbSe		NW, [27] NR [28]
	ZnQ (Q = S, Se)		NW, [29] NR [30]
	CdQ (Q = S, Se, Te)		NW [31–35]
	Reactant	***Product***	
Chemical transformation	Se	M_xSe_y (M = Ag, Pb, Cd)	NW, [13, 36, 37] NT
	Ag_2Se	CdSe	NW [13, 37]
	CdQ (Q=S, Se)	MQ (M = Pb, Zn)	NR [38, 39]
	$LiMo_3Se_3$	Au, Ag, Pt, Pd	NW [40]
	Te	M_xTe_y (M = Ag, Cd, Zn, Pb, Bi, Sb, La, Co)	NW, [41–46] NT, [47–50] NRB [51]
	Te	Ag_2Te–Te–Ag_2Te	NW [52]
	Te	Pt, Pd	NW, [53] NT [53]
	Ag_2Te	MTe (M = Cd, Zn, Pb)	NW [42]
	CdTe	$PtTe_2$	NT [42]
	CdQ (Q = S, Se)	MQ (M = Pb, Zn)	NR [38, 39]
	$Cd(OH)_2$	CdQ (Q = S, Se, Te)	NT [54, 55]

Reproduced by permission of the Royal Society of Chemistry [3]
Abbreviations Nanowire *NW*, nanorod *NR*, nanotube *NT*, nanobelt *NB*, nanoneedle *NN*, nanoribbon *NRB*, and nanowire bundle *NWB*

calculation within generalized gradient approximation (GGA) for exchange-correlation functional, given by Perdew, Burke and Ernzerhof (PBE) as implemented in the Vienna ab initio simulation package (VASP) [58]. The kinetic energy cutoff for the plane wave basis set is set to 500 eV and the core level interactions are represented by the projector augmented wave (PAW) potentials [59]. From our first-principle DFT calculations, we find the surface energy of the basal plane (normal to the *c*-axis) is indeed relatively higher (0.616 J/m^2) than those of the prism planes ($\sim$0.463 J/m^2), clearly reflecting that cutting the stronger covalent bonds in the *c*-direction involves a larger energetic cost to form that surfaces. Due to this difference on surface energies, the calculated theoretical Gibbs-Wulff shape favors a longer *c*-axis than the *a*- and *b*-axis. This corroborates well with the fact that selenium nanowires are reported to show a preferential growth along the [001] direction in a wet chemical process [60]. The 1D growth of this system is also attributed to the

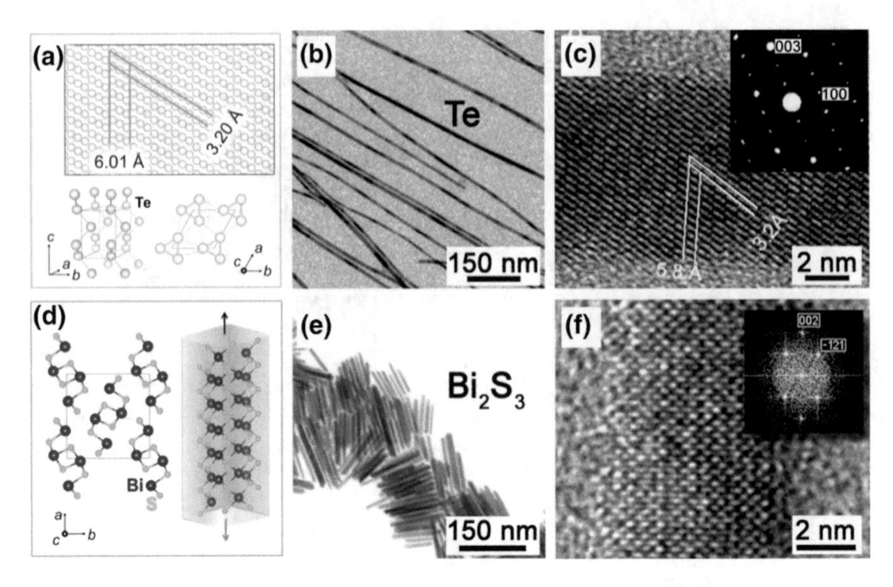

Fig. 2.1 Intrinsically anisotropic growth of 1-D Te, Bi_2S_3 nanocrystals. **a** Tellurium (Te) crystal structure viewed along *a*-axis, while the insets show perspective and top views of the unit cell. Lattice spacings were obtained from the density-functional theory (DFT) calculations with Grimme's van der Waals correction. Projector augmented wavefunction method (PAW) was used for core level potential, while the PBE-GGA exchange correlation functional is used. **b** TEM and **c** HR-TEM images of t-Te nanowires and the electron diffraction (ED) pattern as an inset. Adapted with permission from [14]. Copyright (2006) American Chemical Society. **d** Crystal unit cell of Bi_2S_3 viewed along *c* axis, [001] direction, the growth direction is noted with arrows along *c*-axis, **e** TEM and **f** HR-TEM images of Bi_2S_3 nanorods with an ED pattern (inset in **f**). Reproduced with permission from [64]. Copyright (2007) John Wiley & Sons, Inc.

minimizing of the area of facets (i.e. basal planes) with relatively high surface energies, lowering the total free energy of the system. But in a dry vapor-solid process with relatively high temperature ($\sim$950 °C), the seed effect was found to be more dominant in the growth of nanowires [61]. Yet another study reports the solution-based growth of tellurium 1D crystalline, again attributing to the relatively higher surface energy of the circumferential site of tellurium seed. In this case, it resulted in a tube-like shape [62], where the small energetic difference between central and circumferential site played a crucial role in determining the morphology of the crystal. Interestingly, this small difference in surface energies became negligible when a stabilizing surfactant (PVP) was provided into the system in small portions.

Considerable effort has been devoted to generate diverse 1D Se and Te nanostructures in solution-based synthesis [5–12, 15, 16, 18, 19, 63]. Xia and co-workers reported a sonochemical approach to synthesize Se nanowires [13]. Colloidal spheres of amorphous Se (*a*-Se) were first prepared by reacting selenious acid with aqueous hydrazine as a reducing agent.

After filtering and washing, the *a*-Se colloids were transformed into trigonal Se (*t*-Se) nanowires under mild sonication in ethanol. Resulting Se nanowires were uniform in thickness ($\sim$30 nm) and longer than $\sim$50 μm along the *c*-axis. Yu and co-workers synthesized *t*-Te nanowires 4–9 nm in diameter and hundreds of micrometers in length using hydrothermal methods (Fig. 2.1b, c). [14] HR-TEM images clearly showed that the *t*-Te nanowires grew predominantly along the *c*-axis. Because Se and Te can form a solid solution with the same trigonal crystal structure, binary Se_xTe_y alloys also tend to grow favorably along the *c*-axis, allowing the formation of Se_xTe_y and Se_xTe_y@Te core-shell nanorods [17]. Se and Te satisfy the four requirements of the Hume-Rothery rule for a solid solution between metallic elements: atomic radii difference ($\leq$15%), same crystal structure same valency, and similar electronegativity [65]. The ability to form perfect solid solutions between Se and Te elements indicates that the helical chains of Se_xTe_y nanorods consist of randomly distributed Se and Te along the *c*-axis.

Pnictogen chalcogenides, M_2Q_3 (M = Bi, Sb; Q = S, Se, Te), are nanocrystals with a highly anisotropic crystal structure. Pnictogen chalcogenides have been extensively investigated because of their promising thermoelectric and optoelectronic properties. Interestingly, in addition to their 2D structures, Bi_2S_3, Sb_2S_3, and Sb_2Se_3 possess a strong tendency to grow along the *c*-axis into a 1D nanostructure. Such tendency is attributed to the chain-like molecular conformation that results from V–VI bonding and van der Waals interactions between the chains (Fig. 2.1d). This anisotropic growth mode may, once again, be accounted for on thermodynamic grounds, referring to their surface Gibbs free energy differences between different surfaces. This is illustrated using the orthorhombic Bi_2S_3 as an example. Using first–principles DFT calculations, it was shown that the surface free energy of the (001) surface of Bi_2S_3 (0.423 J/m^2) was higher than the other low-index surfaces [66]. Two other low index facets—(100) and (010) facets (0.359 and 0.349 J/m^2, respectively)—had slightly different surface energies, but were both similarly lower than that of (001). Thus, to afford the minimization of the free energy of the system during growth, Bi_2S_3 is thermodynamically driven to conform to a crystal morphology (i.e. a rod-like shape) which minimizes the area of (001) facet, while maximizing that of the other two facets. A schematic of this equilibrium crystal shape of Bi_2S_3 is shown in Fig. 2.1d.

In numerous experimental studies, Bi_2S_3 nanorods were synthesized through a hot injection method with elemental sulfur, bismuth chloride, and oleylamine [64]. Bismuth chloride dissolved in oleylamine was first transformed into BiOCl, then further reduced into elemental bismuth by the amine groups when the solution was heated up to 170 °C. Uniformly-sized Bi_2S_3 nanorods were produced by rapid hot injection of a sulfur solution. The as-synthesized Bi_2S_3 nanorods were single crystals that showed preferential growth along the [001] direction (Fig. 2.1e, f). If some form of kinetic hindrance occurs during the chemical synthesis process, the final shape of nano crystal may deviate from the ideal one-dimensional morphology, e.g. partially generating quasi-two-dimensional crystals as the kinetic shape. This could then explain the experimental observation in the growth of Bi_2S_3 nanowire, where small portions of nanoplates were grown together with thin-and long nanowires [20]. This delicate difference in the relative surface free energy of the various prism facets of Bi_2S_3 can thus be used to tailor and design the desired morphology for its specific application. Sb_2S_3 and Sb_2Se_3 favorably crystallized into nanorods with an orthorhombic crystal structure [22]. Zheng and co-workers reported the hydrothermal synthesis of 1D Sb_2Se_3 nanostructures with controlled aspect ratios [21]. By adjusting the reaction temperature and concentration of the reactants, Sb_2Se_3 1D nanostructures with different aspect ratios evolved from nanorods to nanowires. Short nanorods were produced at high temperatures (150 °C), which is due to rapid exhaustion of the sources (Sb^{3+} and Se^{2-}) and sequential lack of driving force for nucleation and growth into nanowires. Thick, long nanorods were prepared at high concentrations by continuous reaction between Sb^{3+} and Se^{2-}.

2.1.2 Shape-Guiding Agent Growth

During thermodynamic growth of nuclei, the fastest growing facets should disappear eventually due to the increased surface energy, while the slowest growing facets should survive as the final planes of the product. Controlled growth of nanocrystals with anisotropic morphologies has been extensively investigated by adjusting organic surfactants during the past two decades. Although MC nanoparticles with diverse asymmetric shapes could be experimentally achieved owing to the advances in the synthesis, the interaction between organic molecules and specific facets is still in lack of understanding. This surface energy minimization possibly enables evolution into 1D nanostructure. Specific binding of surfactant molecules to a preferred facet plays a key role in reducing the surface energy. Although Se and Te tend to grow into 1D morphology intrinsically due to their crystal structures, the shapes can vary in synthetic condition governing the preferential growth [68]. Different morphologies of 1D Te nanostructures (NW, NR, NT) were obtained by controlling synthetic parameters such as precursor, surfactant, temperature, pH and reducing agent. Specifically, in the case of Te nanotube, the reduction kinetics can alter Te supersaturation, leading to decrease in diffusion ability of Te atoms and concentration depletion at the surface. Moreover, insufficient coverage of (101) or (102) planes with poly-vinylpyrrolidone (PVP) can induce stacking or aligning these planes on the initial ones during growth [69].

First principles electronic structure simulation revealed the effect of organic surfactant binding on the various surfaces of a CdSe crystal [67, 70]. The typical capping agents include phosphonic acids (PA), phosphine oxides (PO), trimethylamines (TMA), and carboxylic acids (CA). The facets (0001), $(000\bar{1})$, (0110), and $(11\bar{2}0)$ were considered based on the experimental results and the crystal symmetry. Figure 2.2a shows the configurations of the PO and PA bound to the four facets of $Cd_{33}Se_{33}$ dots. The energy was stable when Cd or Se atoms made bonds with the oxygen double bonded to the phosphor (PO and PA) or the carbon (CA). For TMA molecules, the binding took place between Cd or Se atoms and the lone pair

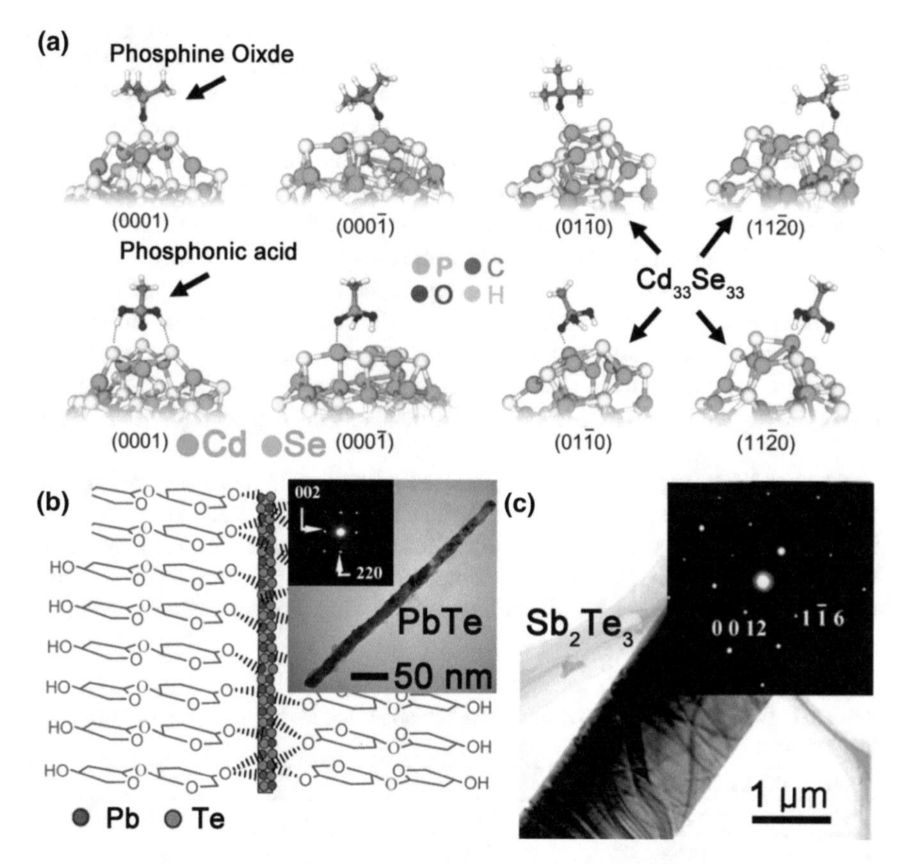

Fig. 2.2 **a** Calculated geometries of phosphine oxide and phosphonic acid bound to the four facets of $Cd_{33}Se_{33}$ cluster. Reproduced with permission from [67]. Copyright (2004) American Chemical Society. **b** Schematic showing the growth mechanism of PbTe nanowires in the presence of sucrose with TEM images and corresponding ED pattern as an inset; anisotropic growth is due to the selective binding between the hydroxyl group sucrose and Pb atoms. Reproduced with permission from [25]. Copyright (2008) American Chemical Society. **c** TEM image of a single Sb_2Te_3 nanobelt with SAED pattern (inset in **c**). Reproduced with permission from [26]. Copyright (2006) American Chemical Society

electrons on the nitrogen atom. Binding energies between the surfactant molecules and the facets of Cd_xSe_x ($x = 15, 33$) were calculated (Table 2.2). PO and PA were found to bind more strongly to the nonpolar side facets with Barvais indices $(01\bar{1}0)$ or $(11\bar{2}0)$ than to the polar surfaces (0001) and $(000\bar{1})$. The preference in binding facets indicates faster growth of polar facets than the nonpolar facets for reducing the total surface energies of the resulting nanostructures. For polar facets, the binding energies of these surfactants to the (0001) surface were much smaller than $(000\bar{1})$, which suggest that the growth of CdSe along the *c*-axis would be dominated by the Se-terminated (0001) facet. In many experimental synthesis, CdSe spheres have been produced in the presence of CA surfactants [71], which is attributed to the small and weak binding energies of CA on the CdSe surfaces. Unlike other surfactants, trimethylamine (TMA) binds slightly stronger to PO and PA were found to bind more strongly to the nonpolar side facets with Barvais indices $(01\bar{1}0)$ or $(11\bar{2}0)$ than to the polar surfaces (0001) and $(000\bar{1})$. The preference in binding facets indicates faster growth of polar facets than the nonpolar facets for reducing the total surface energies of the resulting nanostructures. For polar facets, the binding energies of these surfactants to the (0001) surface were much smaller than $(000\bar{1})$, which suggest that the growth of CdSe along the c-axis would be dominated by the Se-terminated (0001) facet. In many experimental synthesis, CdSe spheres have been produced in the presence of CA surfactants [71], which is attributed to the small and weak binding energies of CA on the CdSe surfaces. Unlike other surfactants, trimethylamine (TMA) binds slightly stronger to the Se-terminated (0001) facets, which can induce the faster growth of Cd-dominated $(000\bar{1})$ along the *c*-axis.

The crystal structures could be engineered by surfactant molecules through fine tuning of surface energies. Generally, wurtzite is a thermodynamically stable crystal structure for both CdS and CdSe in bulk states. The metastable zinc blend phase has been mainly allowed at low reaction temperature (<240 °C) or in small size limit (<4.5 nm). Indeed, zinc blend CdS nanowires with 15 nm diameter could be

Table 2.2 Binding energies (in eV) calculated for various ligands to the surfaces of CdE (E = S, Se) crystals

Wurtzite Cd_xSe_x Ref. [67]				
Ligand	$(000\bar{1})$Cd	(0001)Se	$(01\bar{1}0)$	$(01\bar{2}0)$
PO	1.06a/0.85b	0.66a/0.63b	1.23b	1.37b
PA	1.12a/1.11b	0.66a/0.67b	1.45b	1.26b
CA	0.68a	0.42a	–	–
TMA	0.91a	1.05b	–	–
Wurtzite CdS Ref. [72]				
	(0001)S	(0001)Cd	$(10\bar{1}0)$	$(11\bar{2}0)$
TOP	3.25	0.50	0.50	0.50
OA	3.75	0.50	0.75	2.65
Zinc Blend CdS Ref. [72]				
(001)	(111)S	(111)Cd	(110)	
TOP	3.65	1.30	0.50	0.20
OA	3.20	0.3	0.15	1.60

The energies marked with "a" and "b" were calculated from the $Cd_{15}Se_{15}$ and $Cd_{33}Se_{33}$ clusters, respectively

The elements (Cd, Se, S) placed beside surface notations represent the element terminated at the surfaces. Adapted with permission from [67, 72]. Copyright (2004, 2011) American Chemical Society

Abbreviations of the ligands are as follows: Phosphine Oxide *PO*, Phosphonic Acid *PA*, Carboxylic Acid *CA*, Trimethylamine *TMA*, Trioctylphosphine *TOP*, and Oleic Acid *OA*

successfully synthesized even at high reaction temperature (310 °C). Theoretical analysis confirmed that trioctylphosphine (TOP) molecules are likely to bind strongly to the (001) surface of the zinc blend CdS than to the surfaces of the wurtzite CdS (Table 2.2) [67, 72].

In some cases, a mixture of surfactants can effectively direct 1D growth. Ultrathin (1.7 nm) Hexagonal Cu_2S nanowires as long as tens of micrometers were successfully synthesized using a mixture of dodecanethiol (DT) and oleic acid (OA) as the solvent [23]. DT served as the primary capping ligand for nanowire growth through the formation of thiol-to-Cu(I) bonding on the surface. The growth of Cu_2S nanowires could be attributed to the synergistic effect of the solvent OA and the adsorbent affinity of DT. The energy decrease of the target facet by the organic surfactant is often not large enough to guide the growth to a 1D nanostructure. In this case, use of an organic template whose molecules form complexes with the inorganic sources can be a powerful way to generate anisotropic nanocrystals [24, 25, 73]. Due to the preferential bonding of the organic molecule to a specific inorganic element, the inner surface of the template is covered by the preferred element. The reduced atoms are stabilized by the complex molecules and nuclei are initiated at the inner surface of the template. Consecutive assembly of the inorganic elements in a 1D template facilitates the growth into nanowires.

Ma and co-workers demonstrated that PbTe can be synthesized into nanowires and nanorods in the presence of sucrose templating molecules [25]. Normally, PbTe grows thermodynamically into spherical or cubic nanoparticles due to its preferred isotropic cubic crystal structure. The key factor in promoting the 1D growth was the addition of sucrose, which has hydroxyl groups that interact exclusively with Pb (Fig. 2.2b). In addition, the π–π electron interactions of sucrose served as an organic template for the growth of nanowires. PbTe nanocubes and flower-shaped clusters were synthesized under the same conditions when sucrose was replaced by other typical surfactants such as trioctylphosphine (TOP) and trioctylphosphine oxide (TOPO). Selective interactions between the hydroxyl groups and Pb atoms were confirmed by replacing Pb with Pt in the same synthesis process. Pt source caused the formation of highly agglomerated randomly-shaped PtTe crystals in the same experimental condition. This result suggests the lack of effective interactions between sucrose and both elements (Pt, Te). PbTe nanowires with a thickness of 10–16 nm and length of 400–600 nm were obtained in the presence of sucrose by injecting precursors into pentanediol solvent heated at 210 °C.

This templating approach is a promising way to be applicable diverse MC nanocrystals. However, several factors to determine the shape of the nanocrystals are mingled; the effect of the reduction rate of the precursors, the change of dipole according to the solvent in use, the stability and thermodynamic shape of the template in the synthetic condition, and so on. Hence, weak complexation between the source element and surfactant molecules may not large enough to generate stable templates. In this case, the use of the ionic surfactants can be a noticeable approach to produce anisotropic nanocrystals. The molecules containing ionic charges exist in the forms of vesicles or micelles when their concentration is over a critical value. Zhang and coworkers successfully synthesized Sb_2Te_3 nanobelts with a length of several tens to hundred micrometers in the presence of an anionic surfactant, sodium bis(2-ethylhexyl)sulfosuccinate (AOT) [26]. The pnictogen chalcogenides such as Bi_2Se_3, Bi_2Te_3, and Sb_2Te_3 have been known to preferentially form 2D anisotropic crystal structures. However, the AOT molecules suppressed the growth along the *a*- or *b*-axes of the Sb_2Te_3 crystals and formed 1D templates in which the growth of the crystal is confined. In the absence of AOT, irregular 2D nanoplates were obtained under identical reaction conditions (Fig. 2.2c). Ionic surfactant AOT has been also used for controlled synthesis and evolution of ZnS nanowires with diameters of 30 nm and lengths up to 2.5 μm [74]. Under specific AOT concentration, rod-like micelle of AOT acts as a template for the formation of ZnS nanowires. Zn-thiourea complexes may undergo slow decomposition to produce nanowires in the micelle templates. The same group has also produced ZnSe nanorods using the AOT micelle-template [75].

2.1.3 Axial Oriented Attachment

Classically, crystal has been considered to grow through (i) spontaneous growth of nuclei by consecutive atomic addition or (ii) dissolution of unstable small particles followed by reprecipitation onto more stable particles. The latter concept, the so-called Ostwald ripening process, is based on the solubility difference explained by the Gibbs-Thompson equation [76, 77]. The ripening process has been used to explain particle coarsening, but cannot fully explain the crystal growth behavior or the shape evolution [78]. Since Penn and Banfield first pioneered the oriented attachment as a new crystal growth mechanism [79–81], several anisotropic growth in MC nanocrystals including CdTe, CdSe, CdS, PbSe, ZnS, and ZnSe, could be explained [27–33]. The oriented attachment refers to the direct self-organization of two particles into a single crystal caused by sharing a common crystallographic orientation. A large single crystal can evolve through consecutive attachment of the small nanocrystals, which is different from the simple physical assembly of nanomaterials (which we will discuss later). Because the oriented attachment involves the assembly of nanocrystal building blocks, tailoring the way of attachment generates 1D or 2D structures with tunable properties. The driving force of this attachment is removal of the high-energy surfaces. As particles approach one another in solution, they are assembled by additional energy between the surfactants (van der Waals interactions or dipole interactions).

When the areal density of the capping agent on the higher energy surface is low or negligible, the surfaces meet each other and form long nanowires or nanobelts (Fig. 2.3a). The defects such as misorientation and stacking faults are caused by the direct coagulation of the nanocrystals [82, 83]. After coalescence, the nanocrystals are thought to rearrange into a particle with identical crystallography, which raises the possibility of transformation into a large single crystal [84, 85]. The epitaxial rearrangement takes place through relaxation of the stress caused by misorientation at the attachment interface. However, the growth via the ripening process can occur

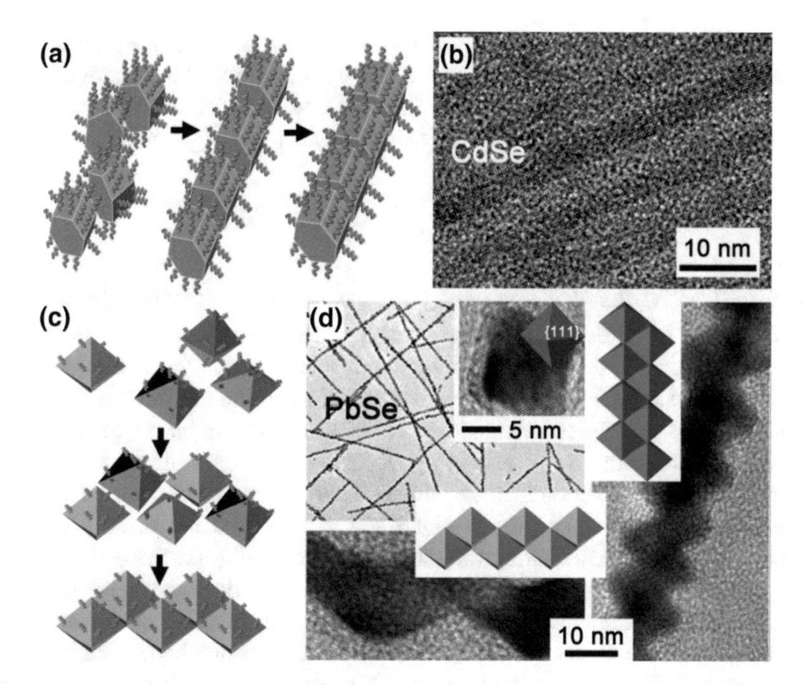

Fig. 2.3 Axial growth by oriented attachment of CdSe, PbSe nanowires. **a** Schematic illustration showing the each growth stage during morphology evolution of CdSe quantum wires. **b** HR-TEM image of CdSe nanowires. Reproduced with permission from [33]. Copyright (2006) American Chemical Society. **c** Schematics of segment showing the assembly and growth of octahedral PbSe nanocrystals into zigzag nanowires. **d** TEM and HR-TEM images of octahedral repeat unit and zigzag nanowires depending on the two attachment mode. (Schemes show the different attachment mode of octahedral PbSe nanocrystals). Adapted with permission from [27]. Copyright (2006) American Chemical Society

simultaneously along orientated attachment. Theoretical and experimental studies have demonstrated that strong surface adsorption of capping ligands to a specific facet can hinder growth via the ripening process and facilitate pure oriented growth [86–88]. Surface adsorption of anions was confirmed to slow down the ripening process because the anions were able to effectively restrict the dissolution of particles in solution [89].

Peng and co-workers have demonstrated the synthesis of CdSe nanowires experimentally and explained theoretically the growth was based on the oriented attachment rather than the continuous axial growth of nanorods (Fig. 2.3a, b) [33, 35]. In their synthesis, CdSe nanowires were synthesized by reacting cadmium acetate ($CdAc_2$) and selenourea in alkylamine solution (octylamine or oleylamine). The synthesis involved three different growth stages: (i) the formation of initial nanocrystals, (ii) prewire aggregates with the configuration of a string of pearls, and (iii) structural transformation into single-crystal nanowires by thermally annealing the prewire aggregates. The prewire aggregates of which the thickness was similar to the diameter of nanocrystals could be interpreted as the key evidence of the oriented attachment. The oriented attachment was described with a thermodynamic model based on the Gibbs free energy. The total free energy G is described in terms of surface energy (γ) weighted by the factors f (such that $\sum_i f_i = 1$),

$$G = \Delta G_f^o + \frac{M}{\rho}(1 - e)[q \sum_i f_i \gamma_i] \quad (2.1)$$

where ΔG_f^o is the standard free energy for the formation of bulk materials, M is molar mass, ρ is the density, q is the surface to volume ratio, and e is the volume dilation induced by the surface stresses σ_i. Details about the assumptions and simplification can be found in the reference [35]. According to the thermodynamic model and first-principles calculations, CdSe basically prefers to grow as short nanorods (aspect ratio: 2–4), indicating that the thermodynamic axial growth by consecutive atomic addition is not a dominant formation mechanism. Furthermore, kinetically controlled axial growth alone was also proved not to be a dominant mechanism from the calculating the nucleation probability on the surfaces of the CdSe nanorods. The growth along <0001> direction via attachment of (0001) facets was the most energetically stable for the growth of CdSe nanowires.

Murray and co-workers synthesized nearly defect-free and highly uniform PbSe nanowires via the oriented attachment [27]. By tailoring reaction conditions, nanowires with diverse shapes were prepared such as straight, zigzag, helical, branched, and tapered nanowires. Attachment of the {100}, {110}, or {111} facets was dependent on the chemical nature of the surfactant(s) used in the synthesis. In the presence of oleic acid only or the co-surfactants of oleic acid and n-tetradecylphosphonic acid (TDPA), nanocrystals assembled along the <100> axis, which was attributed to faster growth of the {111} facet than the {100} facet. Dipole moment along the <100> axis aligned the nanocrystals into nanowires. By replacing TDPA with long, aliphatic primary amines (dodecylamine, hexadecylamine (HDA), oleylamine, etc.), octahedral PbSe nanocrystals containing eight {111} facets formed preferentially due to selective blocking of the {111} facets by the binding of amines. Two types of zigzag nanowires were obtained depending on the attachment mode of the octahedron nanocrystals (Fig. 2.3c, d). Helical nanowires formed when HDA and oleic acid were used as co-surfactants in the reaction medium of trioctylamine. Although PbSe crystals have a highly symmetric cubic structure, oriented attachment leads to successful synthesis of PbSe nanowires with the same crystal structure. The concept of dipole-dipole interactions as a driving force directing 1D nanostructures has been proposed previously [31]. In a subsequent study, Murray and co-workers reported the formation of PbSe/PbS core-shell heterostructures using the pre-formed PbSe nanowires as the building blocks [90].

Highly uniform nanocrystals can assemble to form superstructures with regular particle-particle distances. Self-assembly of the nanocrystals is governed by the attractive forces between the nanocrystals such as van der Waals, Coulombic, and dipole-dipole interactions. Whether the nanocrystals evolve into superstructure by the physical self-assembly or grow to a single crystal by the oriented attachment is currently not in precise control. One of the key factors inducing the oriented attachment is to control the degree of passivation of target surfaces. It requests considerable further understanding about the bonding between organic surfactants and inorganic elements. The attachment of the nanocrystals is also affected by solvent species, reaction temperature, and concentration of nanocrystals. These interactions are often entangled and make it difficult to predict the product. Directed alignment of nanocrystals by external forces may help the oriented attachment process. Alignment of anisotropic nanoparticles is challenging because the interaction between anisotropic nanocrystals is direction-dependent, typically causing raft-like assembly with short range ordering [91–93]. In the presence of an external electric field, the MC nanorods are forced to align along the E-field [94–96]. Russell and co-workers reported 'self-corralling' of CdSe nanorods under an applied electric field [97]. The permanent dipole moment and inherent dielectric property of CdSe nanorods enhanced the alignment of nanorods along their long axis, parallel to the field. This alignment followed by epitaxial merging of the nanocrystals may enable the preparation of long single-crystal nanowires, which may allow the fabrication of aligned nanowires on a substrate.

2.1.4 Chemical Transformation

Chemical transformation from a premade nanocrystal into another is a powerful route to obtain nanocrystals of diverse shapes and chemical compositions that cannot be obtained directly via the conventional synthetic approaches [42, 98]. The synthetic parameters to control the size, shape, and composition of the product nanocrystals are mingled in classical solution-based synthesis. Such multiple control factors interrupt fine-tuning of these variables. However, chemical transformation without any separate homogeneous nucleation allows systematic control over synthetic variables. Chemical transformation has been exploited as a new synthetic tool to generate 1D and 2D nanostructures with more complex morphological features than those obtained by direct synthesis, and has increased access to more arbitrary material systems (Fig. 2.4).

Chemical transformation can be divided into three categories based on the underlying mechanism: alloy formation, ion exchange reaction, and galvanic replacement. Alloy formation is accomplished by mutual diffusion between atoms in premade solids and dissolved atoms in a reduced form. This strategy has been used to generate various MC nanostructures with diverse morphologies including nanoparticles (ZnS, CdS, PbS) [99–101], nanotubes (Ag_2Se, Bi_2Te_3, $CoTe_2$, CoTe) [47–50, 102], nanowires (Ag_2Se, Ag_2Te, Bi_2Te_3, CdTe, PbTe, La_2Te_3 [13, 36, 41–46], heterojunction double dumbbell Ag_2Te–Te–Ag_2Te nanowires [52], and tri-wing Ag_2Te nanoribbons [51]. Chalcogens (Se, Te) are considered an ideal model system to prepare 1D MCs because of their inherent tendency to grow into 1D shapes and their high reactivity with metal precursors. For example, Yu and

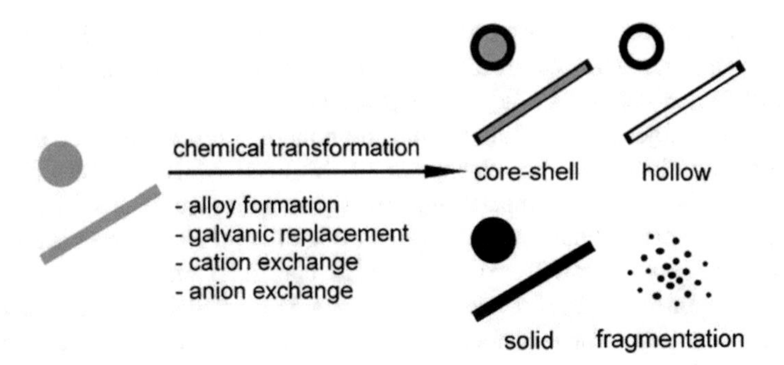

Fig. 2.4 Schematic drawing of chemical transformation of solid nanomaterials. The product may keep their initial template shapes (solid, core-shell, hollow) or break into small fragments. Reproduced with permission from [98]. Copyright (2011) Elsevier Ltd.

co-workers have utilized Te nanowires as starting materials to prepare various 1D metal telluride nanostructures (Bi_2Te_3, CdTe, PbTe) [43, 44]. The reaction processes were simple. Metal cations (Bi^{3+}, Cd^{2+}, Pb^{2+}) were reduced to their neutral elements (Bi, Cd, Pb) using hydrazine hydrate as a reducing agent. The neutral elements reacted with Te nanowires to form metal telluride nanowires. An example of the CdTe system is provided below:

$$2Cd^{2+} + N_2H_4 + 4OH^- \rightarrow 2Cd + N_2 + 4H_2O \quad (2.2)$$

$$Cd + Te(nanowires) \rightarrow CdTe(nanowires) \quad (2.3)$$

Exchange of metal cations in the MC nanocrystals is a useful technique to diversify the accessible material species. Chalcogen anions, which are typically larger than metal cations, play a vital role as a frame preserving the structure of MC materials. Metal cations are mobile within their ionic structure, hence they can be replaced by other cations under appropriate conditions. Numerous research groups have used a solution phase approach to generate 1D chalcogenide nanostructure via cation exchange reactions (CdTe, ZnTe, PbTe, CdSe nanowires, PbS, ZnSe nanorods, $PtTe_2$ nanotubes) [37–39, 42]. Jeong and coworkers systematically demonstrated the chemical transformation of Te nanowires into Ag_2Te nanowires via a topotactic alloying process, and then subsequently conducted cation exchange using the Ag_2Te nanowires to generate diverse MC nanowires (CdTe, ZnTe, PbTe) and $PtTe_2$ nanotubes (Fig. 2.5a).

The as-synthesized Te nanowires with a single-crystal structure (Fig. 2.5b) transformed spontaneously into single crystalline Ag_2Te nanowires with no drastic morphological changes (Fig. 2.5c). This structural preservation can be explained by topotactic lattice matching between trigonal Te and monoclinic Ag_2Te lattices, despite the large volume increase during transformation. Compared to Ag_2Te, other metal tellurides (CdTe, ZnTe, PbTe) are more soluble in polar solvents, which thermodynamically prohibits cation exchange to create other metal telluride

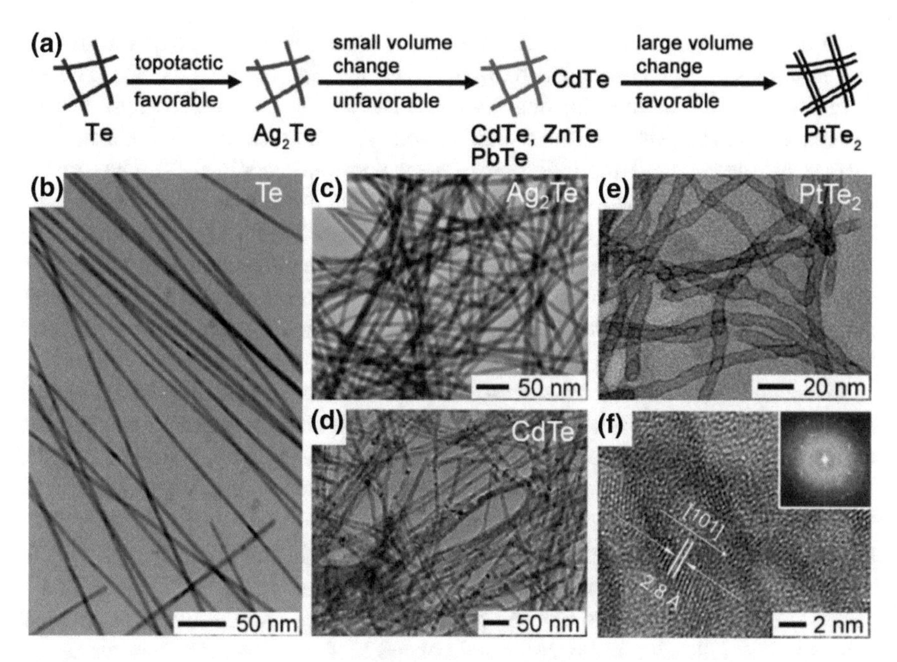

Fig. 2.5 a Flow of chemical transformation from ultrathin Te nanowires into various chalcogenide 1-D nanostructured materials. Topotactic transformation from ultrathin Te to Ag_2Te nanowires, which is thermodynamically favorable. Reversible cation exchange reaction for generating metal telluride nanowires (MTe, M = Cd, Zn, Pb) from Ag_2Te nanowires. The reaction is thermodynamically prohibited, hence the use of specific surfactant is needed. Further transformation of CdTe nanowires into $PtTe_2$ nanotubes through a forward cation exchange reaction. TEM images of ultrathin Te (**b**), Ag_2Te (**c**), CdTe nanowires (**d**). TEM (**e**) and HR-TEM (**f**) images of the $PtTe_2$ nanotubes transformed from CdTe nanowires. The inset shows the Fourier transformed ring pattern of $PtTe_2$ nanotubes. Adapted with permission from [42]. Copyright (2010) American Chemical Society

nanowires from Ag_2Te (Fig. 2.5d). Cation exchange was successfully accomplished by introducing a complexation ligand, tributylphosphine (TBP) which forms selective binding to silver. This example proves that careful choice of ligand can facilitate thermodynamically unfavorable exchanges by reversing the solubility order. Single crystallinity was preserved in the product metal chalcogenide nanowires due to the small volume change. In the transformation of CdTe into $PtTe_2$, cation exchange was thermodynamically favorable due to the much lower solubility of $PtTe_2$ than CdTe. The exchange reaction caused a large volume decrease, resulting in the production of $PtTe_2$ nanotubes (Fig. 2.5e, f).

In contrast to cation exchange, anion exchange causes considerable structural changes. Thus, more energy is required to overcome the kinetic barrier than that required for cation exchange. Anion exchange has rarely been explored compared to cation exchange reactions except a few reports such as CdS, CdSe, CdTe nanowires [103]. CdSe nanotubes with CdSe thorns on their surfaces were synthesized from $Cd(OH)_2$ nanowire bundles through an anion exchange reaction by

Se [55]. $Cd(OH)_2$ nanowire bundles were synthesized first on a glass substrate in aqueous phase. Rapid exchange between OH^- and Se^{2-} as well as fast outward diffusion of Cd^{2+} in the aqueous solution resulted in the formation of tubular structures. Subsequent nucleation and growth of CdSe thorns on the nanotube surface proceeded until the Cd sources were consumed completely. The surface morphology of the nanotubes was dependent on the concentration of the Se ions. Nanotubes with clean and smooth surfaces formed without any thorn at high Se concentrations, which was attributed to uniform chemical reaction along the entire length of the nanowire bundles. Meanwhile, CdSe thorns on the nanotube were generated at low Se concentrations due to localized exchange reactions with Cd^{2+} ions diffusing out from the bundles.

Galvanic replacement is considered a simple, effective way to prepare 1D noble metal nanostructures [104]. The redox potential difference between elements leads to deposition of the more noble elements and dissolution of the less noble elements. Yang and co-workers utilized redox reactions to prepare Au, Ag, Pd, and Pt nanowires from $LiMo_3Se_3$ nanowires; the latter nanowires functioned as both templates and reducing agents [40]. Metal ions ($AuCl_4^-$, Ag^+, $PdCl_4^{2-}$, $PtCl_4^{2-}$) in aqueous solution were reduced and subsequently deposited on the nanowire templates. Meanwhile, the nanowire template was oxidized and dissolved in aqueous solution. The following redox reaction between $LiMo_3Se_3$ and Au took place:

$$3LiMo_3Se_3(s) + AuCl_4^- \rightarrow Au(s) + 3Mo_3Se_3 + 3LiCl + Cl^- \quad (2.4)$$

Pt nanotubes, Pt nanowires, and Pd nanowires a few nanometers in diameter have been successfully prepared by galvanic replacement using Te template nanowires [53]. In a typical synthesis, metal precursors (H_2PtCl_6, $PdCl_2$) are added to an ethylene glycol suspension of Te nanowires at 50 °C. Interestingly, the resulting products had different morphologies. In the case of the Pt nanotubes, some Pt nanoshells formed during the early stages of the reaction, which permitted the inter-diffusion of reactant ions across the shell. As galvanic replacement proceeded, the Pt nanoshells grew inwards at the cost of the Te templates. Based on stoichiometric relationships, this reaction involves the replacement of equivalent molar amounts of Pt by TeO_3^{2-}. The smaller molar volume of Pt ($\sim$9 cm^3/mol) than that of Te ($\sim$20 cm^3/mol) yielded a tubular structure. Furthermore, the molar volume of two moles of Pd ($\sim$18 cm^3/2 mol) is similar to that of 1 mol of Te, which resulted in solid Pd nanowires.

2.2 Synthetic Strategies for 2D Metal Chalcogenides

Exfoliation of graphene from layered graphite triggered other layered ultrathin 2D nanomaterials over the last decade. A family of 2D materials includes single elements (silicene, germanene, phosphorene), boron nitride, metal oxides, and metal chalcogenides. A high mechanical flexibility and transparency of ultrathin 2D nanomaterials are believed to bridge the microscopic properties to the emerging macroscopic features with the quantum confinement from the reduced dimensionality. For example, ultrathin Au nanosheets can be used as a flexible and stretchable electrodes for *p*-type organic semiconductor device. In addition to mechanical exfoliation and CVD, a wet chemistry proved its availability to obtain 2D MC nanocrystals with high quality through liquid exfoliation [105, 106], ion intercalation/exfoliation [107–109], solution-phase deposition [110–113], and chemical transformation [114, 115]. A synthetic protocol for 2D metal chalcogenides has been advancing in a wet chemistry to an extent which one can materialize atomically-thin and wafer-scale MC film for the formation of heterostructures without interlayer contamination [116]. Quite recently, high-performance semiconductor films with atomically-thin nature have been boosting novel integrated circuitry due to its high flexibility and transparency. Solution-based synthetic approaches for such large-scale ultrathin MCs are also

Table 2.3 2D metal chalcogenide nanocrystals reported to date

Mechanism	Product		Morphology
Intrinsic growth	GeQ (Q = S, Se)		NSt [117]
	FeQ, FeQ_x (Q = Se, Te)		NSt, [118] NF [119]
	SnQ (Q = S, Se)		NSt [120, 121]
	MS_2 (M = Ti, Zr, Nb, Ta, Mo), $TiSe_2$		NSt, [122–124] ND [125]
	Bi_2Q_3 (Q = Se, Te)		NP, [126–130] ND, [131] NSt, [131] NF [132]
Shape-guiding agent growth	Te, CdSe		NSt, [133, 134] ND [135]
	$Cu_{2-x}Se$		ND [136]
	In_2S_3		NSt [137]
	$NiTe_2$		NF [138]
Lateral oriented attachment	Bi_2Q_3 (Q = Se, Te)		NP [139]
	CdQ (Q = S, Te)		NSt, [140] NP [141]
	PbQ (Q = S, Te)		NSt [142–144]
	Reactant	***Product***	
Chemical transformation	SnS_2, SnSe	SnTe, MoS_2, $MoSe_2$, WS_2, WSe_2	NSt [113, 114]
	Ti_2S	Cu_2S, CdS	Toroid [145]
	CuS	$CuInS_2$, $CuIn_xGa_{1-x}S_2$, Cu_2ZnSnS_4	NP [146]
	ZnSe	CuSe	Film [147]
	$Co(OH)_2$, $Ni(OH)_2$	Co_3S_4, $CoSe_2$, $CoTe_2$, CdS, NiS	NSt [148]
Ultrathin film growth	MoS_2, [110, 112] $MoSe_2$, [112] WS_2, [110, 112] WSe_2, [112], Sb_2S_3, [149] Sb_2Se_3, [149] $Sb_2(S_{1-x}Se_x)_3$, [149] $SnS_{2-x}Se_x$, [150] $CuInSe_2$, [151] $CuInTe_2$, [152] Cu_2ZnSnS_4, [153] KSb_5S_8 [154]		

Reproduced by permission of the Royal Society of Chemistry [3]
Abbreviations nanoplate *NP*, nanosheet *NSt*, nanodisc *ND*, and nanoflake *NF*

dealt in this chapter. The representative 2D MC nanocrystals are displayed according to their synthetic paths in Table 2.3. As with the synthetic strategies for 1D MC nanocrystals, the following sub-chapters will deal with four different 2D MC synthetic pathways:

1. Intrinsic growth
2. Shape-guiding agent growth
3. Oriented attachment
4. Chemical transformation.

2.2.1 Intrinsic Growth

The shape of nanocrystals is dominated by the surface energy of the facets. For some MCs, 2D shapes (nanoplates, nanodiscs, nanoflakes, nanosheets) are thermodynamically favored. Examples include the VA-VIA (GeS, GeSe, SnS, and SnSe) [117, 120, 121, 155], VIIIB-VIA (FeSe and FeTe) [118, 119], IVB-VIA (TiQ_2, ZrQ_2, and HfQ_2, Q = S, Se, Te) [122, 125, 156–159], VB-VIA ($NbSe_2$ and $TaSe_2$) [125, 160, 161], VIB-VIA (MoS_2, $MoSe_2$, WS_2, and WSe_2) [123, 124, 162], and VA-VIA (Bi_2Se_3, Bi_2Te_3, and Sb_2Te_3) [126, 127, 131, 132, 163, 164]. The structural preferences of the materials listed above have been confirmed by theoretical studies [156, 165] and vapor-phase synthesis [166–168] which has no surfactant-driven effect. The basic crystal structures of the metal chalcogenides are illustrated in Fig. 2.6. They commonly assume the chemical stoichiometries of MX_2 or M_2X_3, where M is the metal and X is the chalcogen. In these structures, the M atoms are often 6-fold coordinated (octahedral distortion occurs when M is a VB or VIB metal atom, see Fig. 2.6a, b, d), while the X atoms are 3-fold coordinated (and also 6-fold for the case of M_2X_3, see Fig. 2.6c). Due to this unique structuring, the 3-fold coordinated X chalcogenide ions form localized pairs of electrons, frequently leading to the formation of van der Waals bonded layering structures, exposing X-terminated layers. Metal chalcogenides may occasionally take on the MX stoichiometry and assume the distorted B1 structure (Fig. 2.6e). These metal chalcogenides are characterized by a MX stoichiometric bi-layered structure, weakly bonded by van der Waals forces.

In solution-based synthesis, the use of surfactants to prevent aggregation of nanocrystals affects the thermodynamic preference and often generates unexpected structures. Jeong and coworkers recently reported a surfactant-free synthesis of Bi_2Te_3 and Bi_2Se_3 nanoflakes in a gram scale (Fig. 2.7) [132]. Formation of the 2D nanocrystals in the absence of any surfactant is a strong evidence of the preference in the growth direction. Bismuth chalcogenide nanoflakes were synthesized by reacting $Bi(NO_3)_3{\cdot}5H_2O$ with Na_2TeO_3 (for Bi_2Te_3) and Na_2SeO_3 (For Bi_2Se_3) in ethylene glycol at 180 °C. The layered crystal structures of bismuth chalcogenides

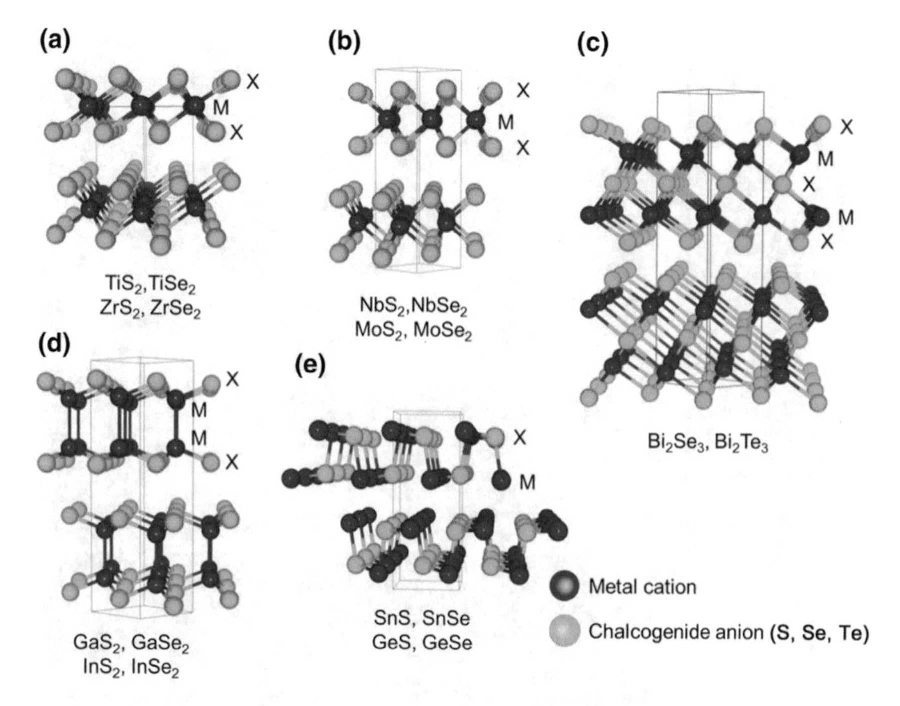

Fig. 2.6 Crystal structures of various MC layered materials: Schematic descriptions of various metal chalcogenide crystals with layered geometry with stoichiometries MX_2 (in **a**, **b**, and **d**), M_2X_3 (in C), and MX (in **e**). Reproduced by permission of the Royal Society of Chemistry [3]

are illustrated in Fig. 2.7a. Five successive atomic planes [$Te_1(Se_1)$–Bi–$Te_2(Se_2)$–Bi–$Te_1(Se_1)$] constitute one quintuple along the *c*-axis. Basically, bismuth chalcogenides preferentially grow into 2D nanostructures because of weak Van der Waals interactions between adjacent $Te_1(Se_1)$ atomic planes that confer bismuth chalcogenides with an intrinsically anisotropic bonding nature. Because the top and bottom surfaces which consist of Te or Se atoms are not stabilized by surfactant, secondary nanoplates grew vertically, as shown in Fig. 2.7b, d. Clear lattice fringes are visible in the HR-TEM images shown in Fig. 2.7c (Bi_2Te_3) and Fig. 2.7e (Bi_2Se_3) and the corresponding selected area electron diffraction (SAED) patterns demonstrate the single crystalline nature of the nanoplates. The cross-section of the Bi_2Se_3 nanocrystals (Fig. 2.7f) indicated 6 quintuples in thickness.

As long as the agglomeration is avoided or the aggregates are redispersed in solution by ultrasonic treatment, synthesis of the surfactant-free 2D nanocrystals deserves further investigation. The layer-structured MCs have a high chance of success to be produced in the absence of surfactant. In most layered MCs, stacking of the atomic layer ends typically with the negatively-charged chalcogen atoms or the positively-charged metal atoms. Depending on the metal species and crystal structure, the uniform direction of the dipoles over the large basal planes provides

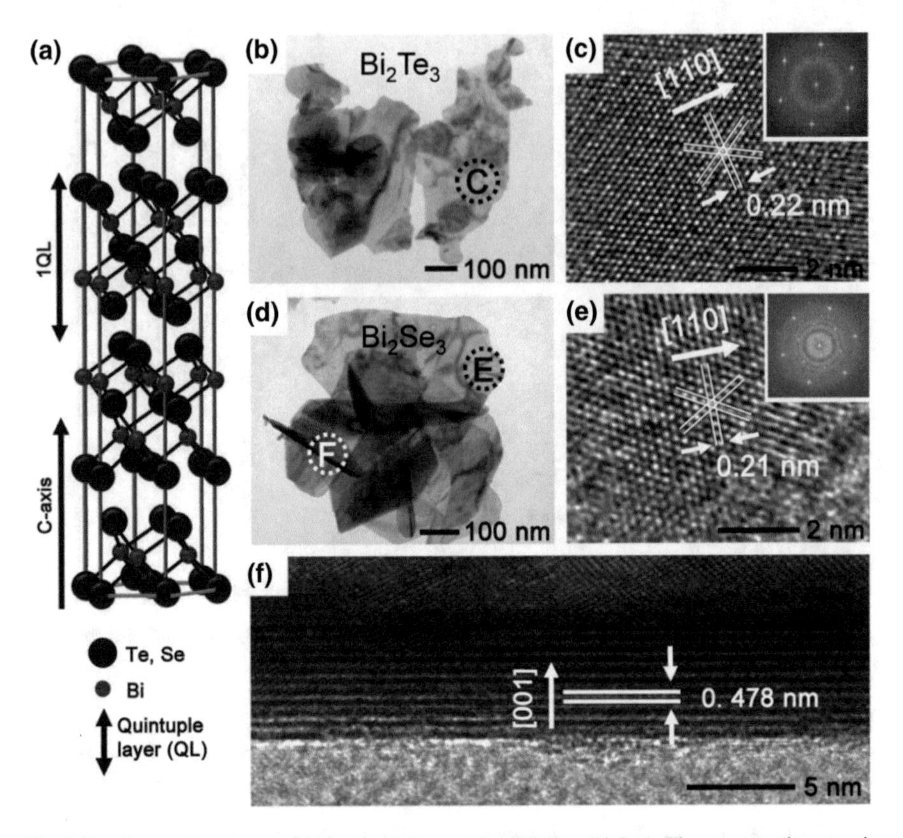

Fig. 2.7 **a** Layered structure of bismuth chalcogenides (Bi_2Te_3, Bi_2Se_3). Five consecutive atomic planes [$Te_1(Se_1)$–Bi–$Te_2(Se_2)$–Bi–$Te_1(Se_1)$] consisting of one quintuple layer along the *c*-axis. **b** TEM image of as-synthesized Bi_2Te_3 nanoflakes and **c** HR-TEM image of the area designated by the letter C in panel B. **d** TEM image of as-synthesized Bi_2Se_3 nanoflakes and **e** A HR-TEM image of the area indicated by the **e** in panel D. **f** HR-TEM image showing a side view of the Bi_2Se_3 nanoflake indicated by the letter F in panel D. The insets in **c** and **d** are the corresponding SAED patterns. Reproduced with permission from [132]. Copyright (2007) John Wiley & Sons, Inc.

better colloidal stability compared to the isotropic particles. Slight amount of inorganic chalcogenide ligands may greatly enhance the colloidal stability in polar solvents. Recently, Talapin and coworkers found that various molecular metal chalcogenide complexes (MCCs) such as $[Sn_2S_6]^{4-}$ and $[Sn_2Se_6]^{4-}$ could serve as convenient ligands for colloidal nanocrystals [136]. The surface inorganic surfactants greatly facilitate charge transport between individual nanocrystals [169]. The insulating organic surfactants of the typical 2D nanocrystals diminish the unique properties of the 2D nanocrystals such as high electron mobility and topological insulator. The inorganic ligands may allow the use of solution-based 2D MCs as the semiconducting active layer for future electronic devices.

2.2.2 *Shape-Guiding Agent Growth*

As in the case of 1D growth, 2D chalcogenide nanostructures can be obtained via surfactant-driven growth. 2D growth of MCs that does not form a layered structure is a challenging subject because of the lack of a driving force for anisotropic growth. In addition to lowering the surface energy of specific facets, surfactants can serve as soft templates for 2D nanocrystals. Despite its strong tendency to grow into 1D nanostructures of Te, 2D Te nanostructure have been developed by using surfactant in liquid phase [134]. Another example includes the hexagonal wurtzite crystals that normally grow isotropically into 0D nanocrystals or sometimes into 1D nanowires because the (0001) facet has the highest surface energy [170]. However, Hyeon and coworkers succeeded to prepare lamellar-structured CdSe nanosheets as thin as 1.4 nm [133]. They employed a soft colloidal template method, in which organic layers consisting of cadmium chloride alkyl amine complexes functioned as 2D templates to generate CdSe nanosheets. In a zinc blend cubic structure, polar axes along the <111> or <001> direction usually facilitate the growth into 1D nanocrystals. Such dipole moments exist through the alternating layers of Cd and Se elements. Recently, Peng and coworkers synthesized CdSe 2D quantum disks (1–3 nm in thickness) with a zinc blend crystal structure [135]. They suppressed potential 1D growth by adopting the polar axes as the short axes of the quantum disks. More specifically, both basal planes with an excess layer of Cd ions were passivated with the negatively charged carboxylate groups of the deprotonated fatty acid ligands, thereby neutralizing the dipole induced by the Cd ions. The close packing of the hydrocarbon chains of the fatty acids enabled the 2D growth of CdSe nanocrystals. Various 2D chalcogenide nanocrystals with non-layered structures have been prepared using a similar synthetic route, including $Cu_{2-x}Se$ nanodiscs [171], In_2S_3 nanosheets [137], and $NiTe_2$ nanoflakes [138]. Dubertret and coworkers extended this strategy to synthesize CdSe/CdS and CdSe/CdZnS core-shell nanoplatelets [172]. They first synthesized zinc blend CdSe nanoplatelets capped by carboxylate ligands [173], and exchanged the ligand by dodecanethiol without disrupting the shape of the thin nanoplatelets. Layer-by-layer deposition of S and Cd enabled CdS coating with a controllable shell-thickness.

Very recently, carbon-coated FeS nanosheets with lateral lengths of 100–200 nm and thicknesses of 4–10 nm have been prepared by surfactant-assisted solution-based synthesis [174]. The governing factor for controlling the morphology of the FeS nanocrystals was the shape of the micelle directed by 1-dodencanethiol (DDT).

DDT was used as both the sulfur source and the surfactant to produce the 2-D FeS nanostructures. At a high concentration of DDT (molar ratio of $Fe(acac)_3$: DDT = 1:20), an Fe/DDT complex with a lamellar structure was formed in oleylamine solution. The complex served as a soft template for the nucleation and growth of the FeS nanosheets. Spherical micelles of Fe/DDT complex were formed at a lower concentration of DDT (1:4), hence polycrystalline nanoparticles of FeS were produced. The hydrocarbon tails of DDT at the surfaces of the nanosheets turned into an amorphous carbon layer after sintering at 400 °C under nitrogen atmosphere, resulting in the formation of carbon-coated FeS nanosheets.

Even in layer-structured systems, surfactants have a critical impact on the final shape of the nanocrystals. The production of uniform-sized 2D nanocrystals with a clean surface is difficult to obtain in the absence of surfactants. The effects of surfactant on the synthesis of ultrathin ($\sim$4 nm) Bi_2Se_3 nanodiscs and nanosheets have recently been investigated (Fig. 2.8) [131]. Bi_2Se_3 grew faster laterally and the basal planes were covered with negatively-charged Se atoms. When negatively charged surfactant, poly(acrylic acid) (PAA), was used, flower-like nanocrystals were obtained, which were similar to the surfactant-free products. In contrast, well-defined Bi_2Se_3 nanodiscs were obtained in the presence of a weakly-positive

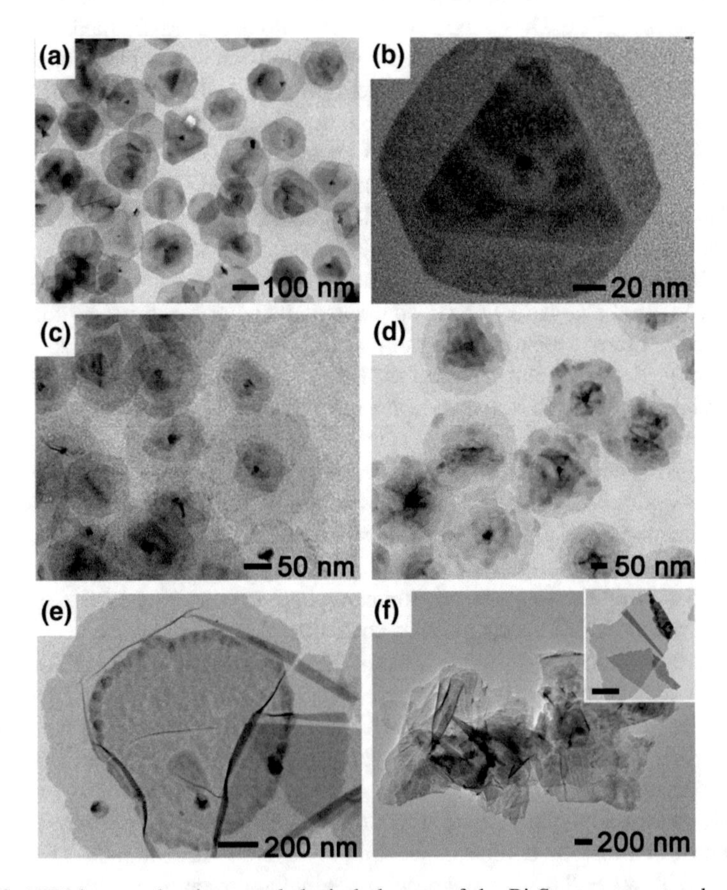

Fig. 2.8 TEM images showing morphological changes of the Bi_2Se_3 nanostructured materials from nanodiscs to nanosheets by controlling the molar ratio of PVP in the mixture with PEI. PVP : PEI = 10:0 (**a**, **b**), 9:1 (**c**), 8:2 (**d**), 3:7 (**e**), and 0:10 (**f**). The inset in F indicates the nanosheet fragments torn off from a wide sheet by applying ultrasonic sound in the suspension (Scale bar = 200 nm). Adapted with permission from [131]. Copyright (2012) American Chemical Society

polymer surfactant, poly(vinylpyrollidone) (PVP). The effect of surfactant charge on the 2D shape formation was examined by mixing PVP and poly(ethylene imine) (PEI) as co-surfactants at different molar ratios: 10:0 (Fig. 2.8a, b), 9:1 (Fig. 2.8c), 8:2 (Fig. 2.8d), 3:7 (Fig. 2.8e), and 0:10 (Fig. 2.8f). Binding of the positively-charged primary amines to the basal plane was much stronger than that of the tertiary amines in PVP. As the molar ratio of PEI increased, the surface energy of the basal plane decreased, resulting in wide nanosheets with maintaining the thickness of the nanodiscs. This effect of surfactant charge was utilized to synthesize various 2D MCs in the form of nanoplates and nanosheets. Sb_2Te_3 hexagonal nanoplates have been synthesized using both solvothermal [163] and hydrothermal approaches [164].

2.2.3 Lateral Oriented Attachment

Although oriented attachment is a relatively new synthetic route, recent advances have proven that this strategy can be used to generate diverse 2D MC nanocrystals [139–144]. Preparation of the 2D MC nanocrystals by the oriented attachment is relatively feasible compared to the formation of 1D nanocrystals by the same approach. Because the layer-structured MCs tend to expose chalcogen atoms on the top and bottom surface of the nanocrystals, selective passivation of those surfaces can be readily achieved even with weakly-binding surfactants. The side surfaces of the thin nanocrystals are passivated loosely compared to the top and bottom surfaces. The side surfaces with a higher energy should be the active sites merging by the oriented attachment. Growth into nanoplates or nanosheets is energetically favorable due to the reduced surface energy in the top and bottom surface. This oriented attachment is expected to be more effective when the thickness of the nanocrystals is small and the binding of surfactants to the top and bottom surface is strong.

The time-dependent shape evolution during the lateral oriented attachment has been observed in several 2D MC nanocrystal system [131, 175]. For example, single-crystalline SnSe nanosheets with a thickness of $\sim$1 nm and a width of $\sim$300 nm are synthesized by heating a mixture solution of SeO_2, $SnCl_4 \cdot 5H_2O$, oleylamine, and 1,10-phenanthroline (Phen) at 120 °C followed aging at 260 °C for 30 min. The TEM studies visualized the growth process which included nucleation of SnSe nanocrystals, aggregation of the nanocrystals into 2D polycrystalline pseudo-sheets, transformation of the pseudo-sheets into single-crystalline nanosheets. This growth mechanism was consistent with the result demonstrated by Schaak and co-workers [121].

In their synthesis, 1,10- phenanthroline (Phen) played a vital role as a capping agent in determining the formation of SnSe nanosheets. In an early stage, Phen bound strongly to the basal plane of the newly formed SnSe nanosheets in cooperation with oleylamine, which suppressed the vertical growth of nanosheets. In the absence of Phen, 3D SnSe nanoflowers were obtained because the stabilization of the basal plane by oleylamine molecules was insufficient to prevent the secondly growth of SnSe nanoplates on the basal planes.

Weller and co-workers utilized this strategy to synthesize PbS nanosheets from PbS nanoparticles (Fig. 2.9) [143, 144]. They based their approach on the standard synthetic protocol used to prepare spherical PbS nanoparticles except that they added chlorine-containing solvents such as 1,2-dichloroethane (DCE) in the oleic acid. The chlorine-containing solvent modified the nucleation and growth rate

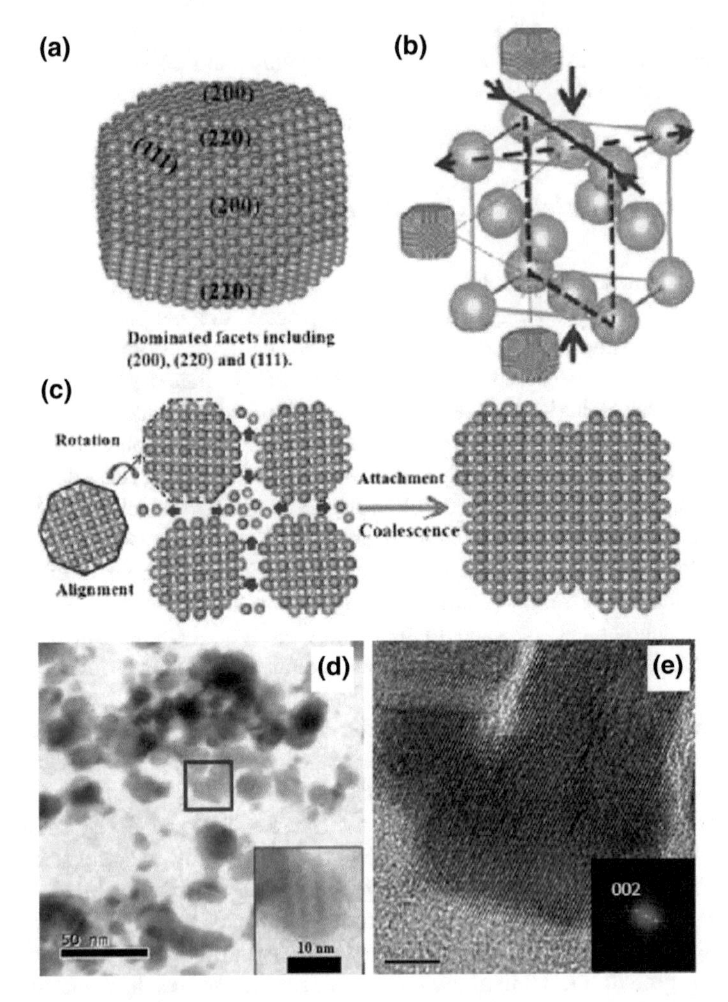

Fig. 2.9 Scheme illustrating the oriented attachment of PbS nanocrsytals into a nanosheet. **a** Reconstructed PbS nanoparticle terminated at surfaces by (200), (220), and (111) facets. **b** Uniaxial compression and resulted compression and tension in a plane to the oriented supercrystal (11). **c** Stress-tuned rotation and alignment along supercrystal (110) and coalesced single crystal sheet via oriented attachments. **d** TEM images of nanosheet structures with a width of 2.5 nm and **e** the enlarged image of the red square. Reproduced with permission from [144]. Copyright (2011) American Chemical Society

during primary nanocrystal formation, which enabled survival of the highly reactive {110} surfaces in small nanocrystals during the early stage of the reaction. These tiny nanocrystals merged into 2D nanosheets because the oleic acid self-assembled monolayer exclusively decreased the surface energy of the {100} facet. Surface stress was suggested as one of the most important feature when inducing the oriented attachment and coalescence of nanoparticles by influencing free movements and interaction of nanoparticles.

2.2.4 Chemical Transformation

Chemical transformation can also be used to obtain 2D nanocrystals with non-layered crystal structures. However, the chemical transformation of 2D nanostructures remains largely unexplored due to its anisotropy. Cheon and coworkers reported the cation exchange reaction of layered TiS_2 nanodiscs to toroidal Cu_2S nanostructures (Fig. 2.10a) [145]. They first synthesized TiS_2 nanodiscs stabilized by oleylamines (Fig. 2.10b–d). Addition of $CuCl_2$ to the nanodisc suspension and subsequent heating at 200 °C for 30 min yielded Cu_2S toroids (Fig. 2.10e–g). The resulting toroids with a hole in their center were highly symmetrical and double-convex. This transformation was driven by a regioselective edge reaction and ion diffusion through nanochannels between layers. The replacement of TiS_2 with Cu ions took place at the edge of the nanodiscs during the early reaction stages, and then subsequent ion diffusion through the interlayer in the TiS_2 nanodiscs resulted in the formation of heteroepitaxial TiS_2–Cu_2S intermediates. As the replacement reaction proceeded, double-convex toroidal Cu_2S structures were generated. The anion exchange reaction was also observed in the strategy for 2D metal chalcogenides. Nickel sulfide nanoplates were transformed from nickel hydroxide nanoplates through anion exchange [9]. SnSe nanosheets with a layered structure were transformed into SnTe nanosheets with a non-layered cubic structure, via solution phase anion exchange [176]. The bond energy difference between trioctylphosphine (TOP) as a chalcogen complex agent and chalcogen elements drove the exchange reaction. Because the P=Se bond is stronger than P=Te, TOP-Se is favorably formed in a reaction system that includes TOP-Te and SnSe, which drives the anion exchange reaction. SnTe nucleated with crystallographic alignment on SnSe nanosheets, which were consumed to yield porous SnTe nanosheets. In 2D nanostructures, Sb_2Te_3 hexagonal nanoplates were transformed into porous 3D network-shaped Te plates. Tartaric acid (TA) in the reaction solution aided the dissolution of Sb^{3+} ions from Sb_2Te_3 by forming $Sb(TA)_x^{3+}$. Simultaneously, Te^{2-} ions were oxidized by O_2 to Te^0; the latter nucleated preferentially on the Sb_2Te_3 surfaces and acted as seeds for growth into porous 3D Te plates. The Te structure was further transformed into porous Pt, Pd, and Au plates via galvanic replacement reactions in the presence of metal precursors (H_2PtCl_6, $PdCl_2$, $HAuCl_4$) [177].

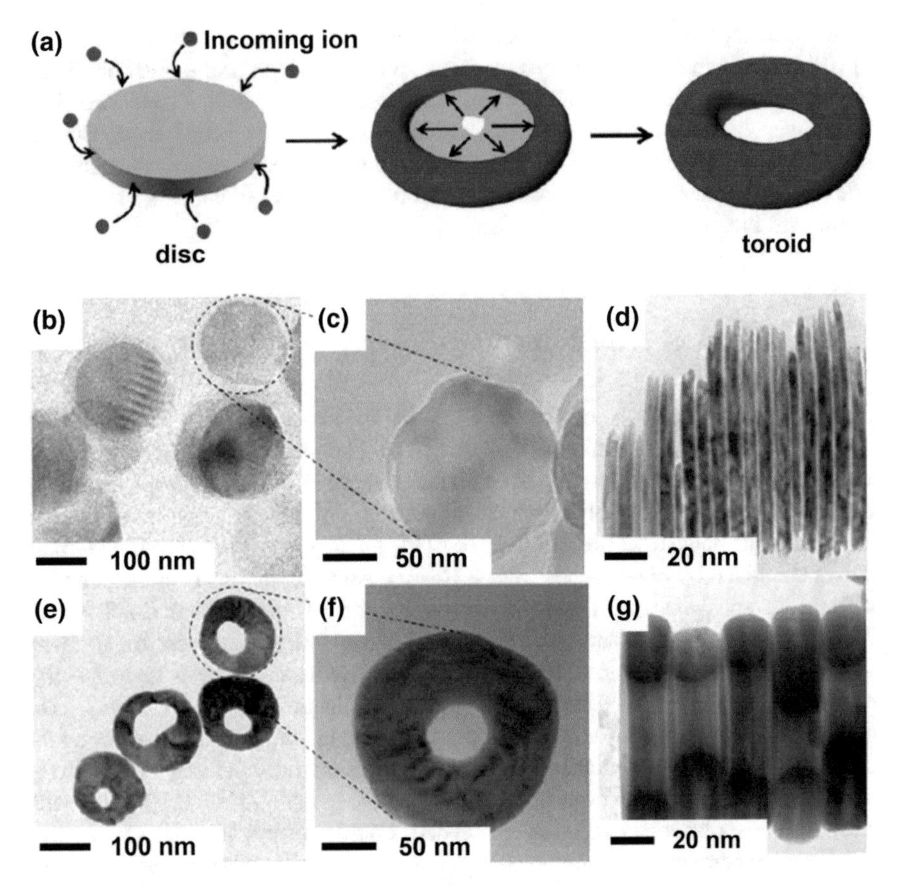

Fig. 2.10 **a** Schematic illustration of the transformation from a Ti_2S nanodisc into a Cu_2S toroid by regioselective reaction of Ti_2S with incoming ions. **b**, **c** TEM images of Ti_2S nanodisc as a starting material. **d** Cross-sectional TEM image of a stack of Ti_2S nanodiscs. **e**, **f** TEM images of Cu_2S toroids transformed from Ti_2S nanodiscs. **g** Side view of a stack of Cu_2S toroids. Reproduced with permission from [145]. Copyright (2011) American Chemical Society

2.3 Assembly Strategies for Anisotropic Metal Chalcogenides

Fabricating ordered arrangement of MCs into 2D or 3D building blocks has advanced in parallel with the development of synthetic route to anisotropic MCs. Individual MC nanocrystal should be incorporated with other components to construct full devices without offsetting the properties of discrete MCs. As a general processing technique, colloidal self-assembly has been, vigorously, studied by exploiting various forces including intermolecular force, electrostatic force, capillary force, and convective motion of solvents [178–180]. The key to self-assembly is that the assembled building blocks are close to or at a thermodynamic equilibrium. Numerous types of self-assembly have been explored to form a hexagonally closely-packed 2D or 3D array for photonic crystals [181], inverse opals [182], chemical sensor [183], and biodegradable scaffolds [184]. The simplest technique includes sedimentation and centrifugation procedures with gravitational and centrifugal forces, respectively [185–187]. Alternative methods were also used to increase the speed of building block formation with enhanced colloidal ordering and control over the thickness: capillary deposition [188], spin-coating [189], electrophoretic assembly [190, 191], convective assembly [192], vertical deposition [193, 194], and other initiative designs [195, 196]. In parallel with the progress of nanocrystal synthesis, positioning and assembly technique have also been developed to fabricate macroscale nanostructures for applications in electronic and energy devices. Monodisperse isotropic nanospheres are well known to easily self-assemble into two- or three-dimensional close-packed superstructures. However, anisotropic nanostructures, such as nanowire and nanosheet, suffer from the difficulty in self-assemble into any ordered and positional superlattice, especially in the case of high aspect ratio. The flexible nature of ultrathin nanowire and nanosheet can cause curling and twisting of individual nanostructures, which makes it difficult to array them in an orientational way.

As a technique for 2D arrays, the Langmuir-Blodgett (LB) process has been widely adopted through solvent evaporation at an air-water interface [197]. The LB technique employs a nonpolar suspension of the particles, typically using hexane or chloroform in order to trap the particles at the interface. A wide variety of nanoscale materials terminated with aliphatic capping agents was assembled into 2D monolayers using the LB process [11, 198–201]. Yu and co-workers have demonstrated that well-defined periodic mesostructures of hydrophilic ultrathin tellurium nanowires were produced through the LB process without surface treatment [11]. Ultrathin Te nanowires with aspect ratio of at least 10^4 can be aligned over a large area and make it possible to construct more complex multilayered structures by repeating transfer. Surface treatment-free LB film of Te nanowires started with suspending Te NW in a mixture solvent of *N,N*-dimethylformamide (DMF) and chloroform. After evaporation of the mixed solvents, the interfacial nanowires shrank to form a closely-packed monolayer due to the interfacial tension [202]. A highly arrayed nanowire film should be adjusted with optimizing the surface pressure for more than 10 h. Figure 2.11a shows the relationship between the surface pressure and surface area (π-A) isotherms of the Te NW monolayer at each stage. The corresponding TEM images are displayed in Fig. 2.11b. At the initial stage (I), the pressure is not enough to form a monolayer film. The nanowires approach each other at small areas just because of the capillary force and van der Walls attraction. At the second stage (II), a higher surface pressure is produced with a decrease of the trough area, indicating the condensed Te nanowire monolayer. At the last stage (III), the surface pressure keeps constant despite a decrease of surface area, which is inferred that loosely arranged nanowires becomes condenser. The slope of π-A isotherms reflects the change rate of surface pressure in terms of how

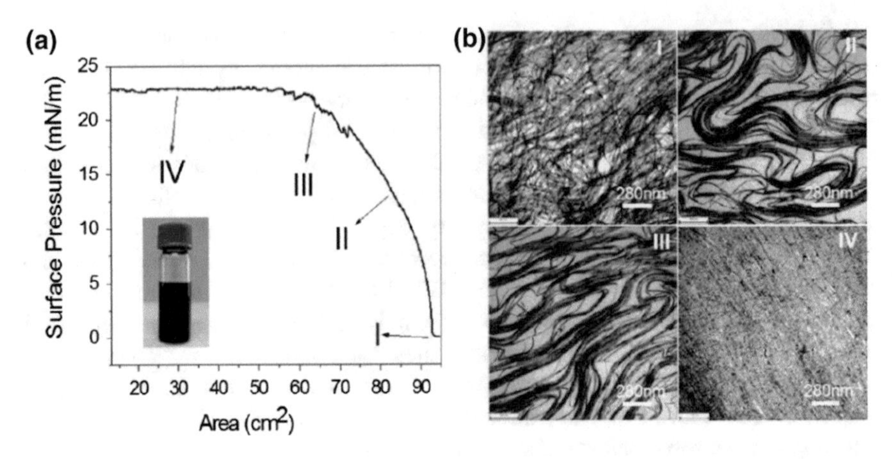

Fig. 2.11 Tellurium nanowire monolayer formation. **a** π-A isotherms of the Te nanowire monolayer at the air/water interface with inset showing Te nanowire dispersion in the mixture of chloroform and DMF. **b** TEM images of the Te nanowire monolayers by compressing with surface pressure 0, 8.5, 19.5, and 23 mN/m, corresponding to I, II, III, IV, respectively. Reproduced with permission from [11]. Copyright (2010) American Chemical Society

nanowires behaves during the monolayer formation. The optimal arrangement can be achieved after duration for 10 h at the surface pressure of 23 mN/m. Thus, the arrayed monolayer film can be broken or folded in parallel with the barrier direction when the surface pressure exceeds the optimal value.

The LB method requires a considerable amount of time to achieve high coverage in the form of monolayer. Instead of LB method, a new approach was suggested to obtain a facile and quick method for a self-assembled monolayers of water-dispersible nanomaterials on water surface [204]. Ultrathin Te nanowires are spread and assembled monolayer film [203]. Well-arrayed Ag_2TeNW film can also be obtained from an aligned Te NW film after self-assembling Te NW into monolayer, followed by transfer to substrates. Ultrathin TeNWs are one of templates to fabricate metal telluride nanowires through chemical transformations. Aqueous synthetic method produced ultrathin Te NWs with a diameter of about 10 nm and a length of several micrometers. The monolayer of Te NWs was made by spreading a TeNW suspension in 1-butanol onto water surface till the formation of robust film (Fig. 2.12b). Te NW films can be transferred to rigid and flexible substrates by contact or scooping and the repeating transfer produces multilayered Te NW films. Once the Te NW film is immersed in an $AgNO_3$ solution in ethylene glycol, the film color changed from dark blue to brown in a matter of seconds, which is indicative of the conversion from Te NW to Ag_2Te NW. Figure 2.12c, d show TEM images of Te NW and Ag_2Te NW films corresponding to the samples on PET substrate (Fig. 2.12e) indicating the periodic array of ultrathin nanowires.

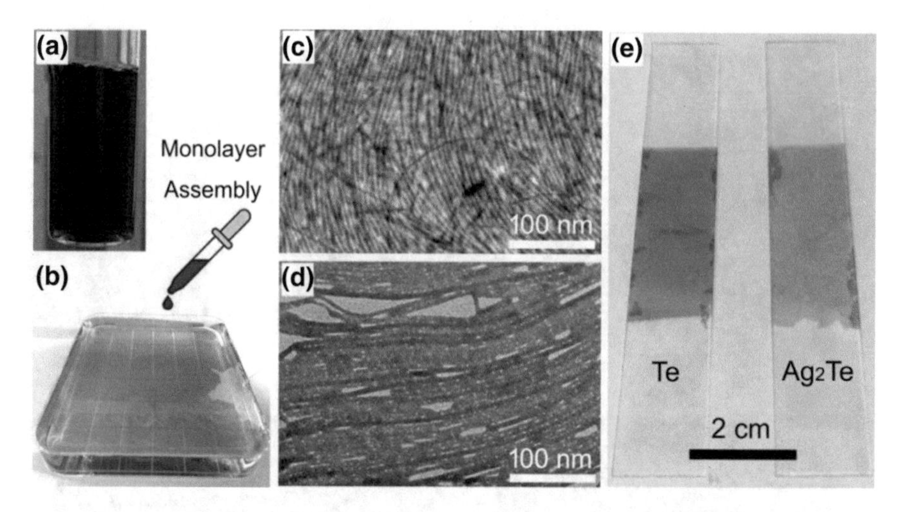

Fig. 2.12 Monolayer assembly of Te nanowires and chemical transformation to Ag_2Te. **a** Te NWs suspended in 1-butanol. **b** Image of Te NW monolayer film on water surface by spreading the Te NW suspension. **c, d** TEM images of Te NW film transferred to a substrate and Ag_2Te NW film transformed from Te NW film. **e** Image of Te NW and Ag_2Te NW film on PET substrates. Reproduced by permission of the Royal Society of Chemistry [203]

PVP-stabilized Te NWs seem to have more space between nanowires than Ag_2Te NW. This may be due to mass transfer along the surface and the large volume change ($\Delta V/V \approx 100\%$) from Te to Ag_2Te after chemical transformation.

Assembly of 1D metal chalcogenides paved its way towards the fabrication of ultrathin nanosheets. Two-dimensional PbS nanosheets can be fabricated through collective coalescence of PbS nanowires (1.8 nm, thickness) at the air-water interface with surface pressure at elevated temperature [205]. The synthesized PbS nanowires are covered with the capping ligand trioctylamine (TOA), determining the shape, size, and assembly pitch of ~2.7 nm. PbS nanowires dispersed in chloroform are spread at the air-water interface and compressed, uniaxially, to yield high density parallel assembly of nanowires (Fig. 2.13a). Figure 2.13b shows parallel 2D supercrystalline PbS nanowire array, which can be used as a starting material for the formation of PbS nanosheets. Afterwards, a gentle heating (50 °C for 90 min) results in ultrathin PbS nanosheet with the release and relocation of TOA from the bottom/side surfaces of the nanowires. The low melting point of TOA (34 °C) is supportive for the coalescence mechanism. The energy involved in the PbS nanosheet formation process can be visualized through coalescence reaction dynamics by measuring the decrease in monolayer area with time at a constant surface pressure at different temperatures (Fig. 2.13c). The fastest drop in area was observed at the temperature of 50 °C.

2D MC nanomaterials can be a template to prepare heterojunction architectures by engineering atomic dislocation and the followed self-assembly. Carroll and co-workers have demonstrated that Sb_2Te_3 nanoplates allows dislocation-driven

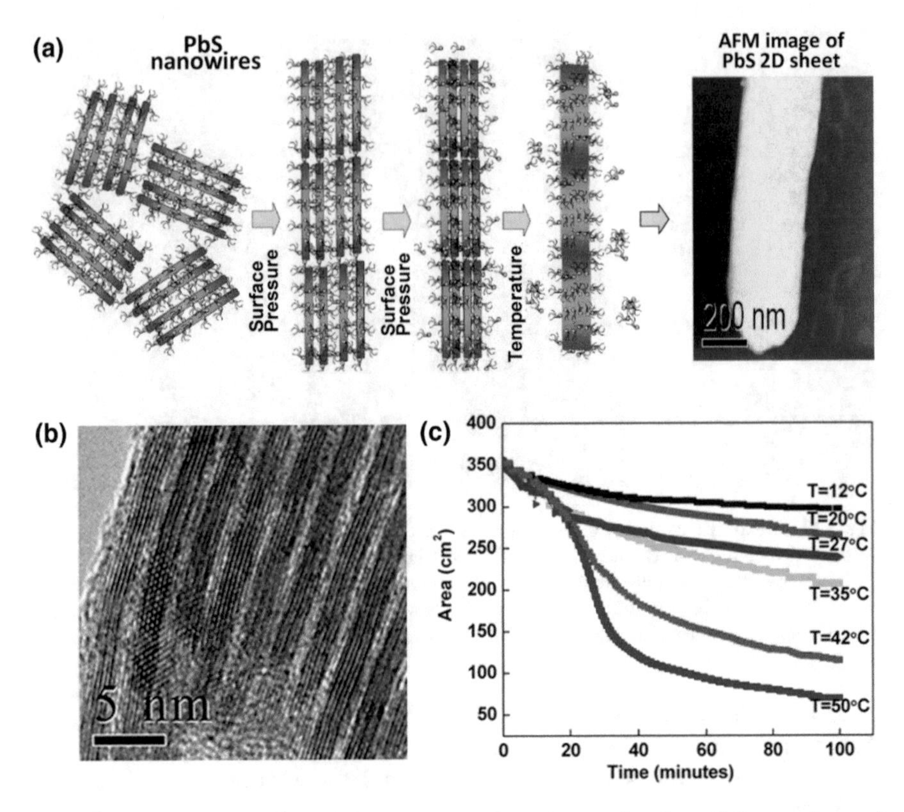

Fig. 2.13 **a** Schematic flow of 2D PbS nanosheet formation starting from PbS nanowires and AFM image of a single sheet showing PbS nanosheet with a width of 200–250 nm along its length. **b** HT-TEM image of PbS nanowires self-assembled into 2D array as a starting material for the nanosheet. **c** Area versus time decay curves at constant surface pressure of 25 mN/m taken at a different temperature range. The decay curves show more surface area loss at higher temperatures. Reproduced with permission from [205]. Copyright (2013) American Chemical Society

self-assembly to form Sb_2Te_3–Ag_2Te–Ag hetero-nanostructures [206]. The growth mechanism of self-assembled Ag nanoparticles at the edge of the Sb_2Te_3 nanoplates is described in Fig. 2.14a. The active Sb_2Te_3 edges with exposed Te dangling bonds are believed to act as heterogeneous nucleation sites. Firstly, These sites react with Ag^+ and then facilitate the growth of Ag nanoparticles. Lateral growth is more preferable (calculated as −0.113 eV/Å^2 Sb_2Te_3(1010)/Ag(111), surface binding energy), so no homogeneous Ag nanoparticles was found in the solution. However, the internal strain of the nanoplates increases with the increased amount of Ag (spiral patterns in Fig. 2.14f, g). The Sb_2Te_3–Ag_2Te–Ag hetero-nanostructures can be stacked into a thin film through vacuum filtration process. Free-standing flexible thin films with ultrathin nature demonstrated its great potential for wearable and self-powered electronic fields.

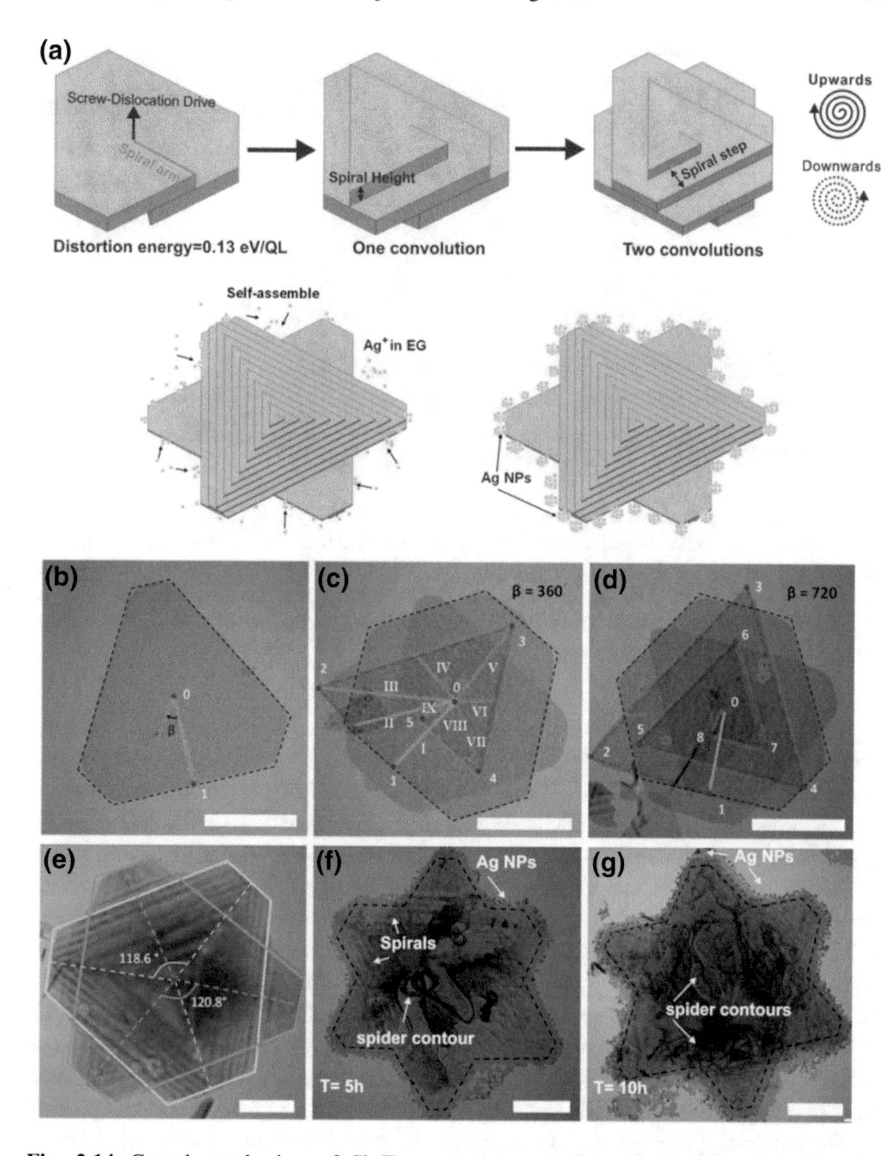

Fig. 2.14 Growth mechanism of Sb_2Te_3 nanoplate and the self-assembling heterojunction architectures. **a** Scheme showing the growth process of spiraled Sb_2Te_3 and Ag–Ag_2Te–Sb_2Te_3 heterojunction through dislocation-driven self-assembly. **b–e** TEM images of Sb_2Te_3 nanoplates at different growth stages, indicating the dislocation generation and formation of bipyramid nanostructures. **f–g** TEM images of self-assembled Ag-decorated nanocomposites after reaction for 5 and 10 h, respectively. Adapted with permission from [206]. Copyright (2017) John Wiley & Sons, Inc.

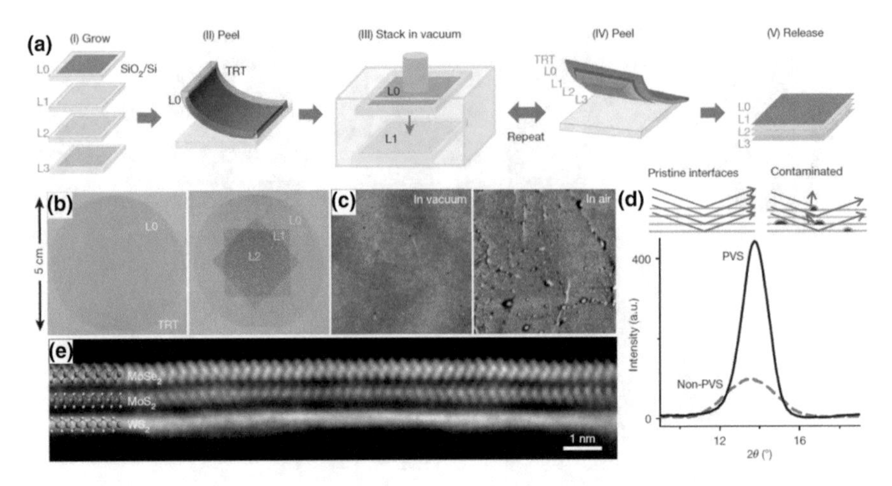

Fig. 2.15 Layer-by-layer assembly of wafer-scale $MoSe_2/MoS_2/WS_2$ thin film. **a** A scheme showing the programmed vacuum stack (PVS) process. **b** Images of wafer-scale MoS_2 films after step (II) and step (IV) in the scheme, respectively. **c** AFM height images obtained from the bottom (L2) side of three-layer MoS_2 stacked in vacuum (left) and air (right). **d** XRD pattern of four-layer MoS_2. **e** Cross-sectional STEM image of a $MoSe_2/MoS_2/WS_2$ film. Reproduced with permission from [116]. Copyright (2017) Springer Nature

Sequential layer-by-layer assembly has been adopted to realize semiconductor films with vertical composition containing atomic-scale precision for modern integrated circuitry [207, 208]. The atomically-thin 2D MC heterostructures include $WS_2/MoSe_2$ [209, 210], graphene/WSe_2 [211], WSe_2/MoS_2 [212], $WSe_2/SnSe_2$ [213], etc. Nevertheless, large-scale assembly is hard to maintain the intrinsic properties of 2D building blocks along the pristine interlayer interfaces due to the previous reports done only by small-scale proof-of-concept demonstrations. To tackle this problem, Park and co-workers have reported the large-scale layer-by-layer assembly process of 2D MC materials into wafer-sized heterostructures [116]. They suggested a so-called 'Programmed vacuum stack (PVS) process' to fabricate wafer-scale $MoSe_2/MoS_2/WS_2$ heterostructured film as follows: (I) synthesis of 2D MC building block, individually, (II) spin-coating with an adhesive polymer film, (III) stacking the first MC layer in a vacuum, (IV) peeling off the substrate, (V) repeating the stacking and peeling-off until the heterostructure MC film is achieved (Fig. 2.15a). The PVS method provides two advantages over MOCVD and etchant/solvent-needed process in terms of large-scale fabrication and generating pristine interlayer interfaces.

Figure 2.15b shows a three-layer MoS_2 film with a 5-cm-diameter, circular region of *L0* after the initial peeling (left) and additional MoS_2 layers, subsequently, stacked on it (right). Thin film X-ray (XRD) confirmed a PVS four-layer MoS_2 film has a peak at $2\theta = 14°$, corresponding to the monolayer spacing along the c-axis direction, the intensity of which is five times higher than the non-PVS-processed films. The peak position indicates and average interlayer spacing of 0.64 nm, which is close to the expected value of 0.65 nm calculated for twisted MoS_2 multilayers [214]. The vacuum-stacked film is much cleaner without any bubble-like features, suggesting the less interlayer contamination in the films (Fig. 2.15c). The PVS process can generate various 2D MC building blocks without the constraints of lattice mismatch or alignment. A cross-sectional STEM image of a $MoSe_2/MoS_2/WS_2$ film shows the armchair axes of the $MoSe_2$ (top) and MoS_2 (middle) layers are parallel to the electron beam unlike that of WS_2. This image tells that the vertical stacking is viable even in the presence of lattice mismatch (lattice constant: $MoSe_2 > MoS_2$, $\sim$4.2%) or interlayer rotation (MoS_2/WS_2) (Fig. 2.15e).

References

1. Venables JA (2003) Introduction to surface and thin film processes. Cambridge University Press, Cambridge
2. Gates B, Mayers B, Cattle B, Xia Y (2002) Synthesis and characterization of uniform nanowires of trigonal selenium. Adv Func Mater 12(3):219–227. https://doi.org/10.1002/1616-3028(200203)12:3%3c219:AID-ADFM219%3e3.0.CO;2-U
3. Min Y, Moon GD, Kim C-E, Lee J-H, Yang H, Soon A, Jeong U (2014) Solution-based synthesis of anisotropic metal chalcogenide nanocrystals and their applications. J Mater Chem C 2(31):6222–6248. https://doi.org/10.1039/C4TC00586D
4. Mayers B, Xia Y (2002) One-dimensional nanostructures of trigonal tellurium with various morphologies can be synthesized using a solution-phase approach. J Mater Chem 12 (6):1875–1881. https://doi.org/10.1039/B201058E
5. Gates B, Yin Y, Xia Y (2000) A solution-phase approach to the synthesis of uniform nanowires of crystalline selenium with lateral dimensions in the range of 10–30 nm. J Am Chem Soc 122(50):12582–12583. https://doi.org/10.1021/ja002608d
6. Lu Q, Gao F, Komarneni S (2004) Biomolecule-assisted reduction in the synthesis of single-crystalline tellurium nanowires. Adv Mater 16(18):1629–1632. https://doi.org/10.1002/adma.200400319
7. Liu Z, Hu Z, Liang J, Li S, Yang Y, Peng S, Qian Y (2004) Size-controlled synthesis and growth mechanism of monodisperse tellurium nanorods by a surfactant-assisted method. Langmuir 20(1):214–218. https://doi.org/10.1021/la035160d
8. Xi B, Xiong S, Fan H, Wang X, Qian Y (2007) Shape-controlled synthesis of tellurium 1D nanostructures via a novel circular transformation mechanism. Cryst Growth Des 7(6):1185–1191. https://doi.org/10.1021/cg060663d
9. Zhang B, Hou W, Ye X, Fu S, Xie Y (2007) 1D tellurium nanostructures: photothermally assisted morphology-controlled synthesis and applications in preparing functional nanoscale materials. Adv Func Mater 17(3):486–492. https://doi.org/10.1002/adfm.200600566
10. Lin Z-H, Yang Z, Chang H-T (2008) Preparation of fluorescent tellurium nanowires at room temperature. Cryst Growth Des 8(1):351–357. https://doi.org/10.1021/cg070357f
11. Liu J-W, Zhu J-H, Zhang C-L, Liang H-W, Yu S-H (2010) Mesostructured assemblies of ultrathin superlong tellurium nanowires and their photoconductivity. J Am Chem Soc 132 (26):8945–8952. https://doi.org/10.1021/ja910871s
12. Qin D, Zhou J, Luo C, Liu Y, Han L, Cao Y (2006) Surfactant-assisted synthesis of size-controlled trigonal Se/Te alloy nanowires. Nanotechnology 17(3):674. https://doi.org/10.1088/0957-4484/17/3/010

13. Jeong U, Xia Y, Yin Y (2005) Large-scale synthesis of single-crystal CdSe nanowires through a cation-exchange route. Chem Phys Lett 416(4):246–250. https://doi.org/10.1016/j.cplett.2005.09.106
14. Qian H-S, Yu S-H, Gong J-Y, Luo L-B, L-f Fei (2006) High-quality luminescent tellurium nanowires of several nanometers in diameter and high aspect ratio synthesized by a poly (Vinyl Pyrrolidone)-assisted hydrothermal process. Langmuir 22(8):3830–3835. https://doi.org/10.1021/la053021l
15. Mo M, Zeng J, Liu X, Yu W, Zhang S, Qian Y (2002) Controlled hydrothermal synthesis of thin single-crystal tellurium nanobelts and nanotubes. Adv Mater 14(22):1658–1662. https://doi.org/10.1002/1521-4095(20021118)14:22%3c1658:AID-ADMA1658%3e3.0.CO;2-2
16. Liu Z, Hu Z, Xie Q, Yang B, Wu J, Qian Y (2003) Surfactant-assisted growth of uniform nanorods of crystalline tellurium. J Mater Chem 13(1):159–162. https://doi.org/10.1039/B208420A
17. Moon GD, Min Y, Ko S, Kim S-W, Ko D-H, Jeong U (2010) Understanding the epitaxial growth of SexTey@Te core − shell nanorods and the generation of periodic defects. ACS Nano 4(12):7283–7292. https://doi.org/10.1021/nn102196r
18. Xiong S, Xi B, Wang W, Wang C, Fei L, Zhou H, Qian Y (2006) The fabrication and characterization of single-crystalline selenium nanoneedles. Cryst Growth Des 6(7):1711–1716. https://doi.org/10.1021/cg060005t
19. Cao XB, Xie Y, Zhang SY, Li FQ (2004) Ultra-thin trigonal selenium nanoribbons developed from series-wound beads. Adv Mater 16(7):649–653. https://doi.org/10.1002/adma.200306317
20. Zhang H, Ji Y, Ma X, Xu J, Yang D (2003) Long Bi_2S_3 nanowires prepared by a simple hydrothermal method. Nanotechnology 14(9):974. https://doi.org/10.1088/0957-4484/14/9/307
21. Ma J, Wang Y, Wang Y, Chen Q, Lian J, Zheng W (2009) Controlled synthesis of one-dimensional Sb_2Se_3 nanostructures and their electrochemical properties. J Phys Chem C 113(31):13588–13592. https://doi.org/10.1021/jp902952k
22. Yu Y, Wang RH, Chen Q, Peng LM (2006) High-quality ultralong Sb_2Se_3 and Sb_2S_3 nanoribbons on a large scale via a simple chemical route. J Phys Chem B 110(27):13415–13419. https://doi.org/10.1021/jp061599d
23. Liu Z, Xu D, Liang J, Shen J, Zhang S, Qian Y (2005) Growth of Cu_2S ultrathin nanowires in a binary surfactant solvent. J Phys Chem B 109(21):10699–10704. https://doi.org/10.1021/jp050332w
24. Lifshitz E, Bashouti M, Kloper V, Kigel A, Eisen MS, Berger S (2003) Synthesis and characterization of PbSe quantum wires, multipods, quantum rods, and cubes. Nano Lett 3 (6):857–862. https://doi.org/10.1021/nl0342085
25. Yan Q, Chen H, Zhou W, Hng HH, Boey FYC, Ma J (2008) A simple chemical approach for PbTe nanowires with enhanced thermoelectric properties. Chem Mater 20(20):6298–6300. https://doi.org/10.1021/cm802104u
26. Shi W, Yu J, Wang H, Zhang H (2006) Hydrothermal synthesis of single-crystalline antimony telluride nanobelts. J Am Chem Soc 128(51):16490–16491. https://doi.org/10.1021/ja066944r
27. Cho K-S, Talapin DV, Gaschler W, Murray CB (2005) Designing PbSe nanowires and nanorings through oriented attachment of nanoparticles. J Am Chem Soc 127(19):7140–7147. https://doi.org/10.1021/ja050107s
28. W-k Koh, Bartnik AC, Wise FW, Murray CB (2010) Synthesis of monodisperse PbSe nanorods: a case for oriented attachment. J Am Chem Soc 132(11):3909–3913. https://doi.org/10.1021/ja9105682
29. Zhu G, Zhang S, Xu Z, Ma J, Shen X (2011) Ultrathin ZnS single crystal nanowires: controlled synthesis and room-temperature ferromagnetism properties. J Am Chem Soc 133 (39):15605–15612. https://doi.org/10.1021/ja2049258

30. Cozzoli PD, Manna L, Curri ML, Kudera S, Giannini C, Striccoli M, Agostiano A (2005) Shape and phase control of colloidal ZnSe nanocrystals. Chem Mater 17(6):1296–1306. https://doi.org/10.1021/cm047874v
31. Tang Z, Kotov NA, Giersig M (2002) Spontaneous organization of single CdTe nanoparticles into luminescent nanowires. Science 297(5579):237–240. https://doi.org/10.1126/science.1072086
32. Jiang F, Liu J, Li Y, Fan L, Ding Y, Li Y (2012) Ultralong CdTe nanowires: catalyst-free synthesis and high-yield transformation into core-shell heterostructures. Adv Func Mater 22 (11):2402–2411. https://doi.org/10.1002/adfm.201102800
33. Pradhan N, Xu H, Peng X (2006) Colloidal CdSe quantum wires by oriented attachment. Nano Lett 6(4):720–724. https://doi.org/10.1021/nl052497m
34. Srivastava BB, Jana S, Sarma DD, Pradhan N (2010) Surface ligand population controlled oriented attachment: a case of CdS nanowires. J Phys Chem Lett 1(13):1932–1935. https://doi.org/10.1021/jz1006077
35. Barnard AS, Xu H, Li X, Pradhan N, Peng X (2006) Modelling the formation of high aspect CdSe quantum wires: axial-growth versus oriented-attachment mechanisms. Nanotechnology 17(22):5707. https://doi.org/10.1088/0957-4484/17/22/029
36. Gates B, Mayers B, Wu Y, Sun Y, Cattle B, Yang P, Xia Y (2002) Synthesis and characterization of crystalline Ag_2Se nanowires through a template-engaged reaction at room temperature. Adv Func Mater 12(10):679–686. https://doi.org/10.1002/1616-3028(20021016)12:10%3c679:AID-ADFM679%3e3.0.CO;2-%23
37. Jeong U, Camargo PHC, Lee YH, Xia Y (2006) Chemical transformation: a powerful route to metal chalcogenide nanowires. J Mater Chem 16(40):3893–3897. https://doi.org/10.1039/B606682H
38. Luther JM, Zheng H, Sadtler B, Alivisatos AP (2009) Synthesis of PbS nanorods and other ionic nanocrystals of complex morphology by sequential cation exchange reactions. J Am Chem Soc 131(46):16851–16857. https://doi.org/10.1021/ja906503w
39. Li H, Zanella M, Genovese A, Povia M, Falqui A, Giannini C, Manna L (2011) Sequential cation exchange in nanocrystals: preservation of crystal phase and formation of metastable phases. Nano Lett 11(11):4964–4970. https://doi.org/10.1021/nl202927a
40. Song JH, Wu Y, Messer B, Kind H, Yang P (2001) Metal nanowire formation using Mo_3Se_3- as reducing and sacrificing templates. J Am Chem Soc 123(42):10397–10398. https://doi.org/10.1021/ja016818h
41. Samal AK, Pradeep T (2009) Room-temperature chemical synthesis of silver telluride nanowires. J Phys Chem C 113(31):13539–13544. https://doi.org/10.1021/jp901953f
42. Moon GD, Ko S, Xia Y, Jeong U (2010) Chemical transformations in ultrathin chalcogenide nanowires. ACS Nano 4(4):2307–2319. https://doi.org/10.1021/nn9018575
43. Wang K, Liang H-W, Yao W-T, Yu S-H (2011) Templating synthesis of uniform Bi_2Te_3 nanowires with high aspect ratio in triethylene glycol (TEG) and their thermoelectric performance. J Mater Chem 21(38):15057–15062. https://doi.org/10.1039/C1JM12384J
44. Liang H-W, Liu S, Wu Q-S, Yu S-H (2009) An efficient templating approach for synthesis of highly uniform CdTe and PbTe nanowires. Inorg Chem 48(11):4927–4933. https://doi.org/10.1021/ic900245w
45. Ga Tai, Zhou B, Guo W (2008) Structural characterization and thermoelectric transport properties of uniform single-crystalline lead telluride nanowires. J Phys Chem C 112 (30):11314–11318. https://doi.org/10.1021/jp8041318
46. Samal AK, Pradeep T (2010) Lanthanum telluride nanowires: formation, doping, and Raman studies. J Phys Chem C 114(13):5871–5878. https://doi.org/10.1021/jp911658k
47. Zhang G, Yu Q, Yao Z, Li X (2009) Large scale highly crystalline Bi_2Te_3 nanotubes through solution phase nanoscale Kirkendall effect fabrication. Chem Commun 17:2317–2319. https://doi.org/10.1039/B822595H
48. Kim SH, Park BK (2010) Effects of Te nanowire microstructure and Bi^{3+} reduction rate on Bi_2Te_3 nanotubes. J Appl Phys 108(10):102808. https://doi.org/10.1063/1.3511689

49. Li J, Tang X, Song L, Zhu Y, Qian Y (2009) From Te nanotubes to $CoTe_2$ nanotubes: a general strategy for the formation of 1D metal telluride nanostructures. J Cryst Growth 311 (20):4467–4472. https://doi.org/10.1016/j.jcrysgro.2009.08.007
50. Fan H, Zhang Y, Zhang M, Wang X, Qian Y (2008) Glucose-assisted synthesis of CoTe nanotubes in situ templated by Te nanorods. Cryst Growth Des 8(8):2838–2841. https://doi.org/10.1021/cg7011364
51. Zhu H, Luo J, Zhang H, Liang J, Rao G, Li J, Liu G, Du Z (2012) Controlled hydrothermal synthesis of tri-wing tellurium nanoribbons and their template reaction. CrystEngComm 14 (1):251–255. https://doi.org/10.1039/C1CE05734K
52. Som A, Pradeep T (2012) Heterojunction double dumb-bell Ag_2Te-Te-Ag_2Te nanowires. Nanoscale 4(15):4537–4543. https://doi.org/10.1039/C2NR30730H
53. Liang HW, Liu S, Gong JY, Wang SB, Wang L, Yu SH (2009) Ultrathin Te nanowires: an excellent platform for controlled synthesis of ultrathin platinum and palladium nanowires/nanotubes with very high aspect ratio. Adv Mater 21(18):1850–1854. https://doi.org/10.1002/adma.200802286
54. Shinde VR, Gujar TP, Noda T, Fujita D, Lokhande CD, Joo O-S (2009) Ultralong cadmium chalcogenide nanotubes from one-dimensional cadmium hydroxide nanowire bundles by soft solution chemistry. J Phys Chem C 113(32):14179–14183. https://doi.org/10.1021/jp904480v
55. Kim JW, Shim H-S, Ko S, Jeong U, Lee C-L, Kim WB (2012) Thorny CdSe nanotubes via an aqueous anion exchange reaction process and their photoelectrochemical applications. J Mater Chem 22(39):20889–20895. https://doi.org/10.1039/C2JM32751A
56. Cao G (2004) Nanostructures & nanomaterials: synthesis, properties & applications. Imperial College Press
57. Grimme S (2006) Semiempirical GGA-type density functional constructed with a long-range dispersion correction. J Comput Chem 27(15):1787–1799. https://doi.org/10.1002/jcc.20495
58. Perdew JP, Burke K, Ernzerhof M (1996) Generalized gradient approximation made simple. Phys Rev Lett 77 (18):3865–3868. https://doi.org/10.1103/physrevlett.77.3865
59. Blöchl PE (1994) Projector augmented-wave method. Phys Rev B 50(24):17953–17979. https://doi.org/10.1103/PhysRevB.50.17953
60. Gao X, Gao T, Zhang L (2003) Solution-solid growth of [small alpha]-monoclinic selenium nanowires at room temperature. J Mater Chem 13(1):6–8. https://doi.org/10.1039/B209399E
61. Ren L, Zhang H, Tan P, Chen Y, Zhang Z, Chang Y, Xu J, Yang F, Yu D (2004) Hexagonal selenium nanowires synthesized via vapor-phase growth. J Phys Chem B 108(15):4627–4630. https://doi.org/10.1021/jp036215n
62. Liu Z, Li S, Yang Y, Hu Z, Peng S, Liang J, Qian Y (2003) Shape-controlled synthesis and growth mechanism of one-dimensional nanostructures of trigonal tellurium. New J Chem 27 (12):1748–1752. https://doi.org/10.1039/B306782C
63. Zhang B, Dai W, Ye X, Hou W, Xie Y (2005) Solution-phase synthesis and electrochemical hydrogen storage of ultra-long single-crystal selenium submicrotubes. J Phys Chem B 109 (48):22830–22835. https://doi.org/10.1021/jp054214k
64. Malakooti R, Cademartiri L, Akçakir Y, Petrov S, Migliori A, Ozin GA (2006) Shape-controlled Bi_2S_3 nanocrystals and their plasma polymerization into flexible films. Adv Mater 18(16):2189–2194. https://doi.org/10.1002/adma.200600460
65. Mizutani U (2008) Hume-Rothery rules for structurally complex alloy phases. CRC Press, London
66. Wang Y, Chen J, Wang P, Chen L, Chen Y-B, Wu L-M (2009) Syntheses, growth mechanism, and optical properties of [001] growing Bi_2S_3 nanorods. J Phys Chem C 113 (36):16009–16014. https://doi.org/10.1021/jp904448k
67. Puzder A, Williamson AJ, Zaitseva N, Galli G, Manna L, Alivisatos AP (2004) The effect of organic ligand binding on the growth of CdSe nanoparticles probed by Ab Initio calculations. Nano Lett 4(12):2361–2365. https://doi.org/10.1021/nl0485861

68. Yuho M, Ho Jun S, Jong-Jin C, Byung-Dong H, Geon Dae M (2018) Dimensional and compositional change of 1D chalcogen nanostructures leading to tunable localized surface plasmon resonances. Nanotechnology 29(34):345603
69. Min-Seok K, Xing-Hua M, Ki-Hyun C, Seung-Yeol J, Kahyun H, Yun-Mo S (2018) A generalized crystallographic description of all tellurium nanostructures. Adv Mater 30 (6):1702701. https://doi.org/10.1002/adma.201702701
70. Manna L, Wang Cingolani R, Alivisatos AP (2005) First-principles modeling of unpassivated and surfactant-passivated bulk facets of Wurtzite CdSe: a model system for studying the anisotropic growth of CdSe nanocrystals. J Phys Chem B 109(13):6183–6192. https://doi.org/10.1021/jp0445573
71. Qu L, Peng ZA, Peng X (2001) Alternative routes toward high quality CdSe nanocrystals. Nano Lett 1(6):333–337. https://doi.org/10.1021/nl0155532
72. Nag A, Hazarika A, Shanavas KV, Sharma SM, Dasgupta I, Sarma DD (2011) Crystal structure engineering by fine-tuning the surface energy: the case of CdE (E=S/Se) nanocrystals. J Phys Chem Lett 2(7):706–712. https://doi.org/10.1021/jz200060a
73. Xue P, Lu R, Li D, Jin M, Tan C, Bao C, Wang Z, Zhao Y (2004) Novel CuS nanofibers using organogel as a template: controlled by binding sites. Langmuir 20(25):11234–11239. https://doi.org/10.1021/la048582b
74. Lv R, Cao C, Zhu H (2004) Synthesis and characterization of ZnS nanowires by AOT micelle-template inducing reaction. Mater Res Bull 39(10):1517–1524. https://doi.org/10.1016/j.materresbull.2004.04.019
75. Lv R, Cao C, Zhai H, Wang D, Liu S, Zhu H (2004) Growth and characterization of single-crystal ZnSe nanorods via surfactant soft-template method. Solid State Commun 130 (3):241–245. https://doi.org/10.1016/j.ssc.2004.01.030
76. Sugimoto T (1987) Preparation of monodispersed colloidal particles. Adv Coll Interface Sci 28:65–108. https://doi.org/10.1016/0001-8686(87)80009-X
77. Mullin JW (2001) Crystallization, 4th edn. Butterworth-Heinemann, Woburn
78. Zhang Q, Liu S-J, Yu S-H (2009) Recent advances in oriented attachment growth and synthesis of functional materials: concept, evidence, mechanism, and future. J Mater Chem 19(2):191–207. https://doi.org/10.1039/B807760F
79. Penn RL, Banfield JF (1998) Oriented attachment and growth, twinning, polytypism, and formation of metastable phases; insights from nanocrystalline TiO_2. Am Miner 83(9–10):1077–1082. https://doi.org/10.2138/am-1998-9-1016
80. Penn RL, Banfield JF (1998) Imperfect oriented attachment: dislocation generation in defect-free nanocrystals. Science 281(5379):969–971. https://doi.org/10.1126/science.281.5379.969
81. Banfield JF, Welch SA, Zhang H, Ebert TT, Penn RL (2000) Aggregation-based crystal growth and microstructure development in natural iron oxyhydroxide biomineralization products. Science 289(5480):751–754. https://doi.org/10.1126/science.289.5480.751
82. Moldovan D, Yamakov V, Wolf D, Phillpot SR (2002) Scaling behavior of grain-rotation-induced grain growth. Phys Rev Lett 89(20):206101. https://doi.org/10.1103/physrevlett.89.206101
83. Leite ER, Giraldi TR, Pontes FM, Longo E, Beltrán A, Andrés J (2003) Crystal growth in colloidal tin oxide nanocrystals induced by coalescence at room temperature. Appl Phys Lett 83(8):1566–1568. https://doi.org/10.1063/1.1605241
84. Zhuang Z, Zhang J, Huang F, Wang Y, Lin Z (2009) Pure multistep oriented attachment growth kinetics of surfactant-free SnO_2 nanocrystals. Phys Chem Chem Phys 11(38):8516–8521. https://doi.org/10.1039/B907967J
85. Zheng H, Smith RK, Y-w Jun, Kisielowski C, Dahmen U, Alivisatos AP (2009) Observation of single colloidal platinum nanocrystal growth trajectories. Science 324(5932):1309–1312. https://doi.org/10.1126/science.1172104
86. Zhang J, Lin Z, Lan Y, Ren G, Chen D, Huang F, Hong M (2006) A multistep oriented attachment kinetics: coarsening of ZnS nanoparticle in concentrated NaOH. J Am Chem Soc 128(39):12981–12987. https://doi.org/10.1021/ja062572a

87. Zhang J, Wang Y, Zheng J, Huang F, Chen D, Lan Y, Ren G, Lin Z, Wang C (2007) Oriented attachment kinetics for ligand capped nanocrystals: coarsening of Thiol-PbS nanoparticles. J Phys Chem B 111(6):1449–1454. https://doi.org/10.1021/jp067040v
88. Huang F, Zhang H, Banfield JF (2003) Two-stage crystal-growth kinetics observed during hydrothermal coarsening of nanocrystalline ZnS. Nano Lett 3(3):373–378. https://doi.org/10.1021/nl025836+
89. Hu Z, Oskam G, Penn RL, Pesika N, Searson PC (2003) The influence of anion on the coarsening kinetics of ZnO nanoparticles. J Phys Chem B 107(14):3124–3130. https://doi.org/10.1021/jp020580h
90. Talapin DV, Yu H, Shevchenko EV, Lobo A, Murray CB (2007) Synthesis of colloidal PbSe/PbS core-shell nanowires and PbS/Au nanowire − nanocrystal heterostructures. J Phys Chem C 111(38):14049–14054. https://doi.org/10.1021/jp074319i
91. Claridge SA, Castleman AW, Khanna SN, Murray CB, Sen A, Weiss PS (2009) Cluster-assembled materials. ACS Nano 3(2):244–255. https://doi.org/10.1021/nn800820e
92. Liu K, Zhao N, Kumacheva E (2011) Self-assembly of inorganic nanorods. Chem Soc Rev 40(2):656–671. https://doi.org/10.1039/C0CS00133C
93. Zanella M, Bertoni G, Franchini IR, Brescia R, Baranov D, Manna L (2011) Assembly of shape-controlled nanocrystals by depletion attraction. Chem Commun 47(1):203–205. https://doi.org/10.1039/C0CC02477E
94. Ahmed S, Ryan KM (2009) Centimetre scale assembly of vertically aligned and close packed semiconductor nanorods from solution. Chem Commun 42:6421–6423. https://doi.org/10.1039/B914478A
95. Ryan KM, Mastroianni A, Stancil KA, Liu H, Alivisatos AP (2006) Electric-field-assisted assembly of perpendicularly oriented nanorod superlattices. Nano Lett 6(7):1479–1482. https://doi.org/10.1021/nl060866o
96. Kelly D, Singh A, Barrett CA, O'Sullivan C, Coughlan C, Laffir FR, O'Dwyer C, Ryan KM (2011) A facile spin-cast route for cation exchange of multilayer perpendicularly-aligned nanorod assemblies. Nanoscale 3(11):4580–4583. https://doi.org/10.1039/C1NR11031D
97. Gupta S, Zhang Q, Emrick T, Russell TP (2006) "Self-Corralling" nanorods under an applied electric field. Nano Lett 6(9):2066–2069. https://doi.org/10.1021/nl061336v
98. Moon GD, Ko S, Min Y, Zeng J, Xia Y, Jeong U (2011) Chemical transformations of nanostructured materials. Nano Today 6(2):186–203. https://doi.org/10.1016/j.nantod.2011.02.006
99. Shao H-F, Qian X-F, Zhu Z-K (2005) The synthesis of ZnS hollow nanospheres with nanoporous shell. J Solid State Chem 178(11):3522–3528. https://doi.org/10.1016/j.jssc.2005.09.007
100. Cabot A, Smith RK, Yin Y, Zheng H, Reinhard BM, Liu H, Alivisatos AP (2008) Sulfidation of Cadmium at the nanoscale. ACS Nano 2(7):1452–1458. https://doi.org/10.1021/nn800270m
101. Wang Y, Cai L, Xia Y (2005) Monodisperse spherical colloids of Pb and their use as chemical templates to produce hollow particles. Adv Mater 17(4):473–477. https://doi.org/10.1002/adma.200401416
102. Bernard Ng CH, Tan H, Fan WY (2006) Formation of Ag_2Se nanotubes and dendrite-like structures from UV irradiation of a CSe_2/Ag colloidal solution. Langmuir 22(23):9712–9717. https://doi.org/10.1021/la061253u
103. Agarwal R, Krook NM, Ren M-L, Tan LZ, Liu W, Rappe AM, Agarwal R (2018) Anion exchange in II–VI semiconducting nanostructures via atomic templating. Nano Lett 18(3):1620–1627. https://doi.org/10.1021/acs.nanolett.7b04424
104. Mohl M, Kumar A, Reddy ALM, Kukovecz A, Konya Z, Kiricsi I, Vajtai R, Ajayan PM (2010) Synthesis of catalytic porous metallic nanorods by galvanic exchange reaction. J Phys Chem C 114(1):389–393. https://doi.org/10.1021/jp9083508
105. Nicolosi V, Chhowalla M, Kanatzidis MG, Strano MS, Coleman JN (2013) Liquid exfoliation of layered materials. Science 340(6139). https://doi.org/10.1126/science.1226419

106. Coleman JN, Lotya M, O'Neill A, Bergin SD, King PJ, Khan U, Young K, Gaucher A, De S, Smith RJ, Shvets IV, Arora SK, Stanton G, Kim H-Y, Lee K, Kim GT, Duesberg GS, Hallam T, Boland JJ, Wang JJ, Donegan JF, Grunlan JC, Moriarty G, Shmeliov A, Nicholls RJ, Perkins JM, Grieveson EM, Theuwissen K, McComb DW, Nellist PD, Nicolosi V (2011) Two-dimensional nanosheets produced by liquid exfoliation of layered materials. Science 331(6017):568–571. https://doi.org/10.1126/science.1194975
107. Dines MB (1975) Lithium intercalation via n-Butyllithium of the layered transition metal dichalcogenides. Mater Res Bull 10(4):287–291. https://doi.org/10.1016/0025-5408(75)90115-4
108. Zeng Z, Yin Z, Huang X, Li H, He Q, Lu G, Boey F, Zhang H (2011) Single-layer semiconducting nanosheets: high-yield preparation and device fabrication. Angew Chem Int Ed 50(47):11093–11097. https://doi.org/10.1002/anie.201106004
109. Zheng J, Zhang H, Dong S, Liu Y, Nai CT, Shin HS, Jeong HY, Liu B, Loh KP (2014) High yield exfoliation of two-dimensional chalcogenides using sodium naphthalenide. Nat Commun 5:2995. https://doi.org/10.1038/ncomms3995
110. Liu K-K, Zhang W, Lee Y-H, Lin Y-C, Chang M-T, Su C-Y, Chang C-S, Li H, Shi Y, Zhang H, Lai C-S, Li L-J (2012) Growth of large-area and highly crystalline MoS_2 thin layers on insulating substrates. Nano Lett 12(3):1538–1544. https://doi.org/10.1021/nl2043612
111. Yang H, Giri A, Moon S, Shin S, Myoung J-M, Jeong U (2017) Highly scalable synthesis of MoS_2 thin films with precise thickness control via polymer-assisted deposition. Chem Mater 29(14):5772–5776. https://doi.org/10.1021/acs.chemmater.7b01605
112. Giri A, Yang H, Thiyagarajan K, Jang W, Myoung JM, Singh R, Soon A, Cho K, Jeong U (2017) One-step solution phase growth of transition metal dichalcogenide thin films directly on solid substrates. Adv Mater 29(26):1700291. https://doi.org/10.1002/adma.201700291
113. Lee J, Pak S, Giraud P, Lee YW, Cho Y, Hong J, Jang AR, Chung HS, Hong WK, Jeong HY, Shin HS, Occhipinti LG, Morris SM, Cha S, Sohn JI, Kim JM (2017) Thermodynamically stable synthesis of large-scale and highly crystalline transition metal dichalcogenide monolayers and their unipolar n–n heterojunction devices. Adv Mater 29 (33):1702206. https://doi.org/10.1002/adma.201702206
114. Chen H, Chen Z, Ge B, Chi Z, Chen H, Wu H, Cao C, Duan X (2017) General strategy for two-dimensional transition metal dichalcogenides by ion exchange. Chem Mater 29 (23):10019–10026. https://doi.org/10.1021/acs.chemmater.7b03523
115. Beberwyck BJ, Surendranath Y, Alivisatos AP (2013) Cation exchange: a versatile tool for nanomaterials synthesis. J Phys Chem C 117(39):19759–19770. https://doi.org/10.1021/jp405989z
116. Kang K, Lee K-H, Han Y, Gao H, Xie S, Muller DA, Park J (2017) Layer-by-layer assembly of two-dimensional materials into wafer-scale heterostructures. Nature 550:229. https://doi.org/10.1038/nature23905
117. Vaughn DD, Patel RJ, Hickner MA, Schaak RE (2010) Single-crystal colloidal nanosheets of GeS and GeSe. J Am Chem Soc 132(43):15170–15172. https://doi.org/10.1021/ja107520b
118. Oyler KD, Ke X, Sines IT, Schiffer P, Schaak RE (2009) Chemical synthesis of two-dimensional iron chalcogenide nanosheets: FeSe, FeTe, Fe(Se, Te), and $FeTe_2$. Chem Mater 21(15):3655–3661. https://doi.org/10.1021/cm901150c
119. Chen L, Zhan H, Yang X, Sun Z, Zhang J, Xu D, Liang C, Wu M, Fang J (2010) Composition and size tailored synthesis of iron selenide nanoflakes. CrystEngComm 12 (12):4386–4391. https://doi.org/10.1039/C005097K
120. Zhang Y, Lu J, Shen S, Xu H, Wang Q (2011) Ultralarge single crystal SnS rectangular nanosheets. Chem Commun 47(18):5226–5228. https://doi.org/10.1039/C0CC05528J
121. Vaughn DD, In S-I, Schaak RE (2011) A precursor-limited nanoparticle coalescence pathway for tuning the thickness of laterally-uniform colloidal nanosheets: the case of SnSe. ACS Nano 5(11):8852–8860. https://doi.org/10.1021/nn203009v

122. Plashnitsa VV, Vietmeyer F, Petchsang N, Tongying P, Kosel TH, Kuno M (2012) Synthetic strategy and structural and optical characterization of thin highly crystalline titanium disulfide nanosheets. J Phys Chem Lett 3(11):1554–1558. https://doi.org/10.1021/jz300487p
123. Radisavljevic B, Radenovic A, Brivio J, Giacometti V, Kis A (2011) Single-layer MoS_2 transistors. Nat Nanotechnol 6:147. https://doi.org/10.1038/nnano.2010.279
124. Lee C, Li Q, Kalb W, Liu X-Z, Berger H, Carpick RW, Hone J (2010) Frictional characteristics of atomically thin sheets. Science 328(5974):76–80. https://doi.org/10.1126/science.1184167
125. Jeong S, Yoo D, J-t Jang, Kim M, Cheon J (2012) Well-defined colloidal 2-D layered transition-metal chalcogenide nanocrystals via generalized synthetic protocols. J Am Chem Soc 134(44):18233–18236. https://doi.org/10.1021/ja3089845
126. Zhang J, Peng Z, Soni A, Zhao Y, Xiong Y, Peng B, Wang J, Dresselhaus MS, Xiong Q (2011) Raman spectroscopy of few-quintuple layer topological insulator Bi_2Se_3 nanoplatelets. Nano Lett 11(6):2407–2414. https://doi.org/10.1021/nl200773n
127. Lu W, Ding Y, Chen Y, Wang ZL, Fang J (2005) Bismuth telluride hexagonal nanoplatelets and their two-step epitaxial growth. J Am Chem Soc 127(28):10112–10116. https://doi.org/10.1021/ja052286j
128. Soni A, Yanyuan Z, Ligen Y, Aik MKK, Dresselhaus MS, Xiong Q (2012) Enhanced thermoelectric properties of solution grown Bi_2Te_3–xSex nanoplatelet composites. Nano Lett 12(3):1203–1209. https://doi.org/10.1021/nl2034859
129. Son JS, Choi MK, Han M-K, Park K, Kim J-Y, Lim SJ, Oh M, Kuk Y, Park C, Kim S-J, Hyeon T (2012) n-type nanostructured thermoelectric materials prepared from chemically synthesized ultrathin Bi_2Te_3 nanoplates. Nano Lett 12(2):640–647. https://doi.org/10.1021/nl203389x
130. Mehta RJ, Zhang Y, Karthik C, Singh B, Siegel RW, Borca-Tasciuc T, Ramanath G (2012) A new class of doped nanobulk high-figure-of-merit thermoelectrics by scalable bottom-up assembly. Nat Mater 11:233. https://doi.org/10.1038/nmat3213
131. Min Y, Moon GD, Kim BS, Lim B, Kim J-S, Kang CY, Jeong U (2012) Quick, controlled synthesis of ultrathin Bi_2Se_3 nanodiscs and nanosheets. J Am Chem Soc 134(6):2872–2875. https://doi.org/10.1021/ja209991z
132. Min Y, Roh JW, Yang H, Park M, Kim SI, Hwang S, Lee SM, Lee KH, Jeong U (2013) Surfactant-free scalable synthesis of Bi_2Te_3 and Bi_2Se_3 nanoflakes and enhanced thermoelectric properties of their nanocomposites. Adv Mater 25(10):1425–1429. https://doi.org/10.1002/adma.201203764
133. Son JS, Wen XD, Joo J, Chae J, Baek Si, Park K, Kim JH, An K, Yu JH, Kwon SG, Choi SH, Wang Z, Kim YW, Kuk Y, Hoffmann R, Hyeon T (2009) Large scale soft colloidal template synthesis of 1.4nm thick CdSe nanosheets. Angew Chem Int Ed 48(37):6861–6864. https://doi.org/10.1002/anie.200902791
134. Wang Y, Qiu G, Wang R, Huang S, Wang Q, Liu Y, Du Y, Goddard WA, Kim MJ, Xu X, Ye PD, Wu W (2018) Field-effect transistors made from solution-grown two-dimensional tellurene. Nat Electron 1(4):228–236. https://doi.org/10.1038/s41928-018-0058-4
135. Li Z, Peng X (2011) Size/shape-controlled synthesis of colloidal CdSe quantum disks: ligand and temperature effects. J Am Chem Soc 133(17):6578–6586. https://doi.org/10.1021/ja108145c
136. Kovalenko MV, Scheele M, Talapin DV (2009) Colloidal nanocrystals with molecular metal chalcogenide surface ligands. Science 324(5933):1417–1420. https://doi.org/10.1126/science.1170524
137. Acharya S, Dutta M, Sarkar S, Basak D, Chakraborty S, Pradhan N (2012) Synthesis of micrometer length indium sulfide nanosheets and study of their dopant induced photoresponse properties. Chem Mater 24(10):1779–1785. https://doi.org/10.1021/cm3003063
138. Jiang L, Zhu Y-J, Cui J-B (2010) Cetyltrimethylammonium bromide assisted self-assembly of $NiTe_2$ nanoflakes: nanoflake arrays and their photoluminescence properties. J Solid State Chem 183(10):2358–2364. https://doi.org/10.1016/j.jssc.2010.08.004

139. Zhang G, Wang W, Lu X, Li X (2009) Solvothermal synthesis of V–VI binary and ternary hexagonal platelets: the oriented attachment mechanism. Cryst Growth Des 9(1):145–150. https://doi.org/10.1021/cg7012528
140. Tang Z, Zhang Z, Wang Y, Glotzer SC, Kotov NA (2006) Self-assembly of CdTe nanocrystals into free-floating sheets. Science 314(5797):274–278. https://doi.org/10.1126/science.1128045
141. Son JS, Park K, Kwon SG, Yang J, Choi MK, Kim J, Yu JH, Joo J, Hyeon T (2012) Dimension-controlled synthesis of CdS nanocrystals: from 0D quantum dots to 2D nanoplates. Small 8(15):2394–2402. https://doi.org/10.1002/smll.201200506
142. Zhu TJ, Chen X, Meng XY, Zhao XB, He J (2010) Anisotropic growth of cubic PbTe nanoparticles to nanosheets: controlled synthesis and growth mechanisms. Cryst Growth Des 10(8):3727–3731. https://doi.org/10.1021/cg100563x
143. Schliehe C, Juarez BH, Pelletier M, Jander S, Greshnykh D, Nagel M, Meyer A, Foerster S, Kornowski A, Klinke C, Weller H (2010) Ultrathin PbS sheets by two-dimensional oriented attachment. Science 329(5991):550–553. https://doi.org/10.1126/science.1188035
144. Wang Z, Schliehe C, Wang T, Nagaoka Y, Cao YC, Bassett WA, Wu H, Fan H, Weller H (2011) Deviatoric stress driven formation of large single-crystal PbS nanosheet from nanoparticles and in situ monitoring of oriented attachment. J Am Chem Soc 133 (37):14484–14487. https://doi.org/10.1021/ja204310b
145. Jeong S, Han JH, J-t Jang, J-w Seo, Kim J-G, Cheon J (2011) Transformative two-dimensional layered nanocrystals. J Am Chem Soc 133(37):14500–14503. https://doi.org/10.1021/ja2049594
146. Wu XJ, Huang X, Qi X, Li H, Li B, Zhang H (2014) Copper-based ternary and quaternary semiconductor nanoplates: templated synthesis, characterization, and photoelectrochemical properties. Angew Chem Int Ed 53(34):8929–8933. https://doi.org/10.1002/anie.201403655
147. Estrada CA, Zingaro RA, Meyers EA, Nair PK, Nair MTS (1994) Modification of chemically deposited ZnSe thin films by ion exchange reaction with copper ions in solution. Thin Solid Films 247(2):208–212. https://doi.org/10.1016/0040-6090(94)90801-X
148. Zhao W, Zhang C, Geng F, Zhuo S, Zhang B (2014) Nanoporous hollow transition metal chalcogenide nanosheets synthesized via the anion-exchange reaction of metal hydroxides with chalcogenide ions. ACS Nano 8(10):10909–10919. https://doi.org/10.1021/nn504755x
149. Yang B, Xue D-J, Leng M, Zhong J, Wang L, Song H, Zhou Y, Tang J (2015) Hydrazine solution processed Sb_2S_3, Sb_2Se_3 and Sb2($S_1 - xSe_x$)3 film: molecular precursor identification, film fabrication and band gap tuning. Sci Rep 5:10978. https://doi.org/10.1038/srep10978
150. Mitzi DB, Kosbar LL, Murray CE, Copel M, Afzali A (2004) High-mobility ultrathin semiconducting films prepared by spin coating. Nature 428:299. https://doi.org/10.1038/nature02389
151. Milliron DJ, Mitzi DB, Copel M, Murray CE (2006) Solution-processed metal chalcogenide films for p-type transistors. Chem Mater 18(3):587–590. https://doi.org/10.1021/cm052300r
152. Mitzi DB, Copel M, Murray CE (2006) High-mobility p-type transistor based on a spin-coated metal telluride semiconductor. Adv Mater 18(18):2448–2452. https://doi.org/10.1002/adma.200600157
153. Yang W, Duan H-S, Cha KC, Hsu C-J, Hsu W-C, Zhou H, Bob B, Yang Y (2013) Molecular solution approach to synthesize electronic quality Cu_2ZnSnS_4 thin films. J Am Chem Soc 135(18):6915–6920. https://doi.org/10.1021/ja312678c
154. Mitzi DB, Raoux S, Schrott AG, Copel M, Kellock A, Jordan-Sweet J (2006) Solution-based processing of the phase-change material KSb5S8. Chem Mater 18(26):6278–6282. https://doi.org/10.1021/cm0619510
155. Kang J-G, Park J-G, Kim D-W (2010) Superior rate capabilities of SnS nanosheet electrodes for Li ion batteries. Electrochem Commun 12(2):307–310. https://doi.org/10.1016/j.elecom.2009.12.025

156. Wilson JA, Yoffe AD (1969) The transition metal dichalcogenides discussion and interpretation of the observed optical, electrical and structural properties. Adv Phys 18 (73):193–335. https://doi.org/10.1080/00018736900101307
157. Jaegermann W, Tributsch H (1988) Interfacial properties of semiconducting transition metal chalcogenides. Prog Surf Sci 29(1):1–167. https://doi.org/10.1016/0079-6816(88)90015-9
158. Arnaud Y, Chevreton M (1981) Etude comparative des composés TiX_2 (X=S, Se, Te). Structures de TiTe2 et TiSeTe. J Solid State Chem 39(2):230–239. https://doi.org/10.1016/0022-4596(81)90336-4
159. Jang J-T, Jeong S, Seo J-W, Kim M-C, Sim E, Oh Y, Nam S, Park B, Cheon J (2011) Ultrathin Zirconium Disulfide Nanodiscs. J Am Chem Soc 133(20):7636–7639. https://doi.org/10.1021/ja200400n
160. Coleman RV, Giambattista B, Hansma PK, Johnson A, McNairy WW, Slough CG (1988) Scanning tunnelling microscopy of charge-density waves in transition metal chalcogenides. Adv Phys 37(6):559–644. https://doi.org/10.1080/00018738800101439
161. Friend RH, Jerome D (1979) Periodic lattice distortions and charge density waves in one- and two-dimensional metals. J Phys C: Solid State Phys 12(8):1441. https://doi.org/10.1088/0022-3719/12/8/009
162. Rapoport L, Bilik Y, Feldman Y, Homyonfer M, Cohen SR, Tenne R (1997) Hollow nanoparticles of WS2 as potential solid-state lubricants. Nature 387:791. https://doi.org/10.1038/42910
163. Wang W, Poudel B, Yang J, Wang DZ, Ren ZF (2005) High-yield synthesis of single-crystalline antimony telluride hexagonal nanoplates using a solvothermal approach. J Am Chem Soc 127(40):13792–13793. https://doi.org/10.1021/ja054861p
164. Shi W, Zhou L, Song S, Yang J, Zhang H (2008) Hydrothermal synthesis and thermoelectric transport properties of impurity-free antimony telluride hexagonal nanoplates. Adv Mater 20 (10):1892–1897. https://doi.org/10.1002/adma.200702003
165. Hulliger F (1976) Structural chemistry of layer-type phases. Reidel publishing, Dordrecht
166. Takemura Y, Suto H, Honda N, Kakuno K, Saito K (1997) Characterization of FeSe thin films prepared on GaAs substrate by selenization technique. J Appl Phys 81(8):5177–5179. https://doi.org/10.1063/1.365162
167. Wu XJ, Zhang ZZ, Zhang JY, Ju ZG, Li BH, Li BS, Shan CX, Zhao DX, Yao B, Shen DZ (2008) Growth of FeSe on general substrates by metal-organic chemical vapor deposition and the application in magnet tunnel junction devices. Thin Solid Films 516(18):6116–6119. https://doi.org/10.1016/j.tsf.2007.11.012
168. Malik MA, Afzaal M, O'Brien P (2010) Precursor chemistry for main group elements in semiconducting materials. Chem Rev 110(7):4417–4446. https://doi.org/10.1021/cr900406f
169. Liu W, Lee J-S, Talapin DV (2013) III–V nanocrystals capped with molecular metal chalcogenide ligands: high electron mobility and ambipolar photoresponse. J Am Chem Soc 135(4):1349–1357. https://doi.org/10.1021/ja308200f
170. Peng X, Manna L, Yang W, Wickham J, Scher E, Kadavanich A, Alivisatos AP (2000) Shape control of CdSe nanocrystals. Nature 404:59. https://doi.org/10.1038/35003535
171. Choi J, Kang N, Yang HY, Kim HJ, Son SU (2010) Colloidal synthesis of cubic-phase copper selenide nanodiscs and their optoelectronic properties. Chem Mater 22(12):3586–3588. https://doi.org/10.1021/cm100902f
172. Mahler B, Nadal B, Bouet C, Patriarche G, Dubertret B (2012) Core/shell colloidal semiconductor nanoplatelets. J Am Chem Soc 134(45):18591–18598. https://doi.org/10.1021/ja307944d
173. Ithurria S, Bousquet G, Dubertret B (2011) Continuous transition from 3D to 1D confinement observed during the formation of CdSe nanoplatelets. J Am Chem Soc 133 (9):3070–3077. https://doi.org/10.1021/ja110046d
174. Xu C, Zeng Y, Rui X, Xiao N, Zhu J, Zhang W, Chen J, Liu W, Tan H, Hng HH, Yan Q (2012) Controlled soft-template synthesis of ultrathin C@FeS nanosheets with high-li-storage performance. ACS Nano 6(6):4713–4721. https://doi.org/10.1021/nn2045714

175. Li L, Chen Z, Hu Y, Wang X, Zhang T, Chen W, Wang Q (2013) Single-layer single-crystalline SnSe nanosheets. J Am Chem Soc 135(4):1213–1216. https://doi.org/10.1021/ja3108017
176. Sines IT, Vaughn DD, Biacchi AJ, Kingsley CE, Popczun EJ, Schaak RE (2012) Engineering porosity into single-crystal colloidal nanosheets using epitaxial nucleation and chalcogenide anion exchange reactions: the conversion of SnSe to SnTe. Chem Mater 24 (15):3088–3093. https://doi.org/10.1021/cm301734b
177. Zhang H, Wang H, Xu Y, Zhuo S, Yu Y, Zhang B (2012) Conversion of Sb_2Te_3 hexagonal nanoplates into three-dimensional porous single-crystal-like network-structured te plates using oxygen and tartaric acid. Angew Chem Int Ed 51(6):1459–1463. https://doi.org/10.1002/anie.201107460
178. Whitesides GM, Grzybowski B (2002) Self-assembly at all scales. Science 295(5564):2418–2421. https://doi.org/10.1126/science.1070821
179. Whitesides GM, Boncheva M (2002) Beyond molecules: self-assembly of mesoscopic and macroscopic components. Proc Natl Acad Sci 99(8):4769–4774. https://doi.org/10.1073/pnas.082065899
180. Glotzer SC, Solomon MJ (2007) Anisotropy of building blocks and their assembly into complex structures. Nat Mater 6:557. https://doi.org/10.1038/nmat1949
181. Dinsmore AD, Crocker JC, Yodh AG (1998) Self-assembly of colloidal crystals. Curr Opin Colloid Interface Sci 3(1):5–11. https://doi.org/10.1016/S1359-0294(98)80035-6
182. Xia Y, Gates B, Yin Y, Lu Y (2000) Monodispersed colloidal spheres: old materials with new applications. Adv Mater 12(10):693–713. https://doi.org/10.1002/(SICI)1521-4095(200005)12:10%3c693:AID-ADMA693%3e3.0.CO;2-J
183. Holtz JH, Asher SA (1997) Polymerized colloidal crystal hydrogel films as intelligent chemical sensing materials. Nature 389:829. https://doi.org/10.1038/39834
184. Zhang YS, Yao J, Wang LV, Xia Y (2014) Fabrication of cell patches using biodegradable scaffolds with a hexagonal array of interconnected pores (SHAIPs). Polymer 55(1):445–452. https://doi.org/10.1016/j.polymer.2013.06.019
185. Pusey PN, van Megen W (1986) Phase behaviour of concentrated suspensions of nearly hard colloidal spheres. Nature 320:340. https://doi.org/10.1038/320340a0
186. Davis KE, Russel WB, Glantschnig WJ (1989) Disorder-to-order transition in settling suspensions of colloidal silica: X-ray measurements. Science 245(4917):507–510. https://doi.org/10.1126/science.245.4917.507
187. Ballato J, James A (1999) A ceramic photonic crystal temperature sensor. J Am Ceram Soc 82(8):2273–2275. https://doi.org/10.1111/j.1151-2916.1999.tb02078.x
188. Li H-L, Marlow F (2006) Solvent effects in colloidal crystal deposition. Chem Mater 18 (7):1803–1810. https://doi.org/10.1021/cm052294z
189. Jiang P, McFarland MJ (2004) Large-scale fabrication of wafer-size colloidal crystals, macroporous polymers and nanocomposites by spin-coating. J Am Chem Soc 126 (42):13778–13786. https://doi.org/10.1021/ja0470923
190. Trau M, Saville DA, Aksay IA (1996) Field-induced layering of colloidal crystals. Science 272(5262):706–709. https://doi.org/10.1126/science.272.5262.706
191. Rogach AL, Kotov NA, Koktysh DS, Ostrander JW, Ragoisha GA (2000) Electrophoretic deposition of latex-based 3D colloidal photonic crystals: a technique for rapid production of high-quality opals. Chem Mater 12(9):2721–2726. https://doi.org/10.1021/cm000274l
192. Wostyn K, Zhao Y, Yee B, Clays K, Persoons A, Schaetzen Gd, Hellemans L (2003) Optical properties and orientation of arrays of polystyrene spheres deposited using convective self-assembly. J Chem Phys 118(23):10752–10757. https://doi.org/10.1063/1.1573173
193. Jiang P, Bertone J, Hwang K, Colvin V (1999) Single-crystal colloidal multilayers of controlled thickness. Chem Mater 11(8):2132–2140. https://doi.org/10.1021/cm990080+
194. Li J, Han Y (2006) Optical intensity gradient by colloidal photonic crystals with a graded thickness distribution. Langmuir 22(4):1885–1890. https://doi.org/10.1021/la052699y

195. Park SH, Xia Y (1999) Assembly of mesoscale particles over large areas and its application in fabricating tunable optical filters. Langmuir 15(1):266–273. https://doi.org/10.1021/la980658e
196. Wong S, Kitaev V, Ozin GA (2003) Colloidal crystal films: advances in universality and perfection. J Am Chem Soc 125(50):15589–15598. https://doi.org/10.1021/ja0379969
197. Pelton RH, Chibante P (1986) Preparation of aqueous lattices with N-isopropylacrylamide. Colloids Surf 20(3):247–256. https://doi.org/10.1016/0166-6622(86)80274-8
198. Dabbousi BO, Murray CB, Rubner MF, Bawendi MG (1994) Langmuir-Blodgett manipulation of size-selected CdSe nanocrystallites. Chem Mater 6(2):216–219. https://doi.org/10.1021/cm00038a020
199. Collier CP, Saykally RJ, Shiang JJ, Henrichs SE, Heath JR (1997) Reversible tuning of silver quantum dot monolayers through the metal-insulator transition. Science 277 (5334):1978–1981. https://doi.org/10.1126/science.277.5334.1978
200. Fried T, Shemer G, Markovich G (2001) Ordered two-dimensional arrays of ferrite nanoparticles. Adv Mater 13(15):1158–1161. https://doi.org/10.1002/1521-4095(200108)13:15%3c1158:AID-ADMA1158%3e3.0.CO;2-6
201. Tao A, Kim F, Hess C, Goldberger J, He R, Sun Y, Xia Y, Yang P (2003) Langmuir–Blodgett Silver nanowire monolayers for molecular sensing using surface-enhanced Raman spectroscopy. Nano Lett 3(9):1229–1233. https://doi.org/10.1021/nl0344209
202. Li YJ, Huang WJ, Sun SG (2006) A universal approach for the self-assembly of hydrophilic nanoparticles into ordered monolayer films at a toluene/water interface. Angew Chem Int Ed 45(16):2537–2539. https://doi.org/10.1002/anie.200504595
203. Seo HJ, Jeong W, Lee S, Moon GD (2018) Ultrathin silver telluride nanowire films and gold nanosheet electrodes for a flexible resistive switching device. Nanoscale 10(12):5424–5430. https://doi.org/10.1039/C8NR01429A
204. Moon GD, Lee TI, Kim B, Chae G, Kim J, Kim S, Myoung J-M, Jeong U (2011) Assembled monolayers of hydrophilic particles on water surfaces. ACS Nano 5(11):8600–8612. https://doi.org/10.1021/nn202733f
205. Acharya S, Das B, Thupakula U, Ariga K, Sarma DD, Israelachvili J, Golan Y (2013) A bottom-up approach toward fabrication of ultrathin PbS sheets. Nano Lett 13(2):409–415. https://doi.org/10.1021/nl303568d
206. Dun C, Hewitt CA, Li Q, Xu J, Schall DC, Lee H, Jiang Q, Carroll DL (2017) 2D chalcogenide nanoplate assemblies for thermoelectric applications. Adv Mater 29 (21):1700070. https://doi.org/10.1002/adma.201700070
207. Geim AK, Grigorieva IV (2013) Van der Waals heterostructures. Nature 499:419. https://doi.org/10.1038/nature12385
208. Novoselov KS, Mishchenko A, Carvalho A, Castro Neto AH (2016) 2D materials and van der Waals heterostructures. Science 353(6298). https://doi.org/10.1126/science.aac9439
209. Xu W, Liu W, Schmidt JF, Zhao W, Lu X, Raab T, Diederichs C, Gao W, Seletskiy DV, Xiong Q (2016) Correlated fluorescence blinking in two-dimensional semiconductor heterostructures. Nature 541:62. https://doi.org/10.1038/nature20601
210. Rivera P, Seyler KL, Yu H, Schaibley JR, Yan J, Mandrus DG, Yao W, Xu X (2016) Valley-polarized exciton dynamics in a 2D semiconductor heterostructure. Science 351 (6274):688–691. https://doi.org/10.1126/science.aac7820
211. Lin Y-C, Chang C-YS, Ghosh RK, Li J, Zhu H, Addou R, Diaconescu B, Ohta T, Peng X, Lu N, Kim MJ, Robinson JT, Wallace RM, Mayer TS, Datta S, Li L-J, Robinson JA (2014) Atomically thin heterostructures based on single-layer Tungsten Diselenide and graphene. Nano Lett 14(12):6936–6941. https://doi.org/10.1021/nl503144a
212. Wierzbowski J, Klein J, Sigger F, Straubinger C, Kremser M, Taniguchi T, Watanabe K, Wurstbauer U, Holleitner AW, Kaniber M, Müller K, Finley JJ (2017) Direct exciton emission from atomically thin transition metal dichalcogenide heterostructures near the lifetime limit. Sci Rep 7(1):12383. https://doi.org/10.1038/s41598-017-09739-4

213. Roy T, Tosun M, Hettick M, Ahn GH, Hu C, Javey A (2016) 2D-2D tunneling field-effect transistors using $WSe_2/SnSe_2$ heterostructures. Appl Phys Lett 108(8):083111. https://doi.org/10.1063/1.4942647
214. Liu K, Zhang L, Cao T, Jin C, Qiu D, Zhou Q, Zettl A, Yang P, Louie SG, Wang F (2014) Evolution of interlayer coupling in twisted molybdenum disulfide bilayers. Nat Commun 5:4966. https://doi.org/10.1038/ncomms5966

Chapter 3
Applications

Abstract Interest in anisotropic 1D and 2D nanostructures has been steadily increasing due to their large surface area, quantum confinement effect, and superior optoelectrical or thermoelectrical properties. Anisotropic geometry and size reduction comparable to the Debye length can alter the optical, electric, and magnetic properties of the bulk counterparts. Focusing on MCs prepared in solution phases, the major interests for practical applications are thermoelectric power generation or electronic device cooling, high performance electrodes for batteries, nanocrystal-based photovoltaic devices, and photodetectors. Recently, new potential uses of the solution-based MC nanocrystals are being investigated, which include localized surface plasmon resonance (LSPR) and oxygen reduction reaction (ORR). MC thin films prepared by the vacuum process have shown the possibilities as new class of materials such as topological insulator, semiconductor with a high electron mobility, and superconductor. Although the organic surfactants indispensable in solution chemistry diminish such superior physical properties, solution-based approach has large room to achieve such physical properties. This chapter will introduce recent advances in the practical applications of the anisotropic MC nanocrystals, and introduce briefly the potential applications.

3.1 Energy Storage

Graphite, which is conventionally used as the anode of lithium ion batteries (LIBs), has a small theoretical capacity of 372 $m A h g^{-1}$, which is too low to meet the current need in electronic devices. A number of 2D metal disulfide (MS_2) nanocrystals such as MoS_2, WS_2, and SnS_2 are considered as promising alternatives to graphite [1–4]. Their theoretical capacities are typically twice that of graphite and their layered structures (S-M-S) facilitate reversible intercalation of Li ions (Fig. 3.1a). The synthesis of the metal disulfides has been actively developed during the last five years. Hydrothermal and solvothermal methods have been widely used for the synthesis of the 2D MS nanocrystals and their hybrid composite materials with graphene or CNTs [4, 5]. These methods facilitate the large scale production of

G. D. Moon, *Anisotropic Metal Chalcogenide Nanomaterials*,
SpringerBriefs in Materials, https://doi.org/10.1007/978-3-030-03943-1_3

the high quality 2D MS nanocrystals at low cost and in a short processing time. Thermal decomposition approach is promising to prepare the stoichiometric 2D MS_2 nanocrystals with massive production for industrial needs [3]. Free standing 2D MoS_2 or WS_2 nanosheets were synthesized by decomposition of single-source precursors containing metal and sulfur sources in oleylamine which could cover the surfaces of the 2D MS nanosheets. This oleylamine molecule functions as a protective layer for the oxidation and aggregation [6]. Exfoliation of bulk materials in a liquid phase is a relatively new approach to prepare thin 2D nanosheets [7, 8]. Ultrasonic treatment in organic solvents helped the exfoliation of the bulk materials. The surface energy of the bulk material in the solvent should be minimized to lower the energy requested for exfoliation, hence the surface energy of the solvent should be similar to that of the bulk material.

With thermal decomposition method, Chen and co-workers reported that LIBs with anode electrodes fabricated with 2D SnS_2 nanoplates showed enhanced capacities [3]. The discharge capacity was as high as 1311 mA h g^{-1} in the first cycle, which is close to the sum of the theoretical irreversible capacity (587 m A h g^{-1}) and maximum theoretical reversible capacity (645 m A h g^{-1}). The average discharge capacity (583 m A h g^{-1}), close to 90% of the maximum theoretical reversible value, was stable and reversible up to 30 cycles. These excellent electrochemical properties were attributed to enhanced diffusion kinetics of lithium ions by the finite lateral size and open edges of the nanoplates. Recently, the same group synthesized ZrS_2 nanodiscs with tunable lateral dimensions (20, 35, and 60 nm) and systematically studied the effect of size on Li ion intercalation (Fig. 3.1b, c) [9]. The average discharge capacities of the 20, 35, and 60 nm ZrS_2 nanodiscs were 586, 527, and 433 m A h g^{-1}, respectively (Fig. 3.1d). The 20 nm nanodiscs had a capacity 2.3 times greater than that of bulk ZrS_2 material (255 m A h g^{-1}). The retention capacities of the small nanoplates were ~80, ~77, and ~71% of the original capacities, respectively, while the retention of bulk ZrS_2 decreased continuously and was only 39% after 50 cycles (Fig. 3.1e).

MS_2/carbon composite nanocrystals such as FeS/C [11], SnS_2/graphene [12], CoS_2/C [13], MoS_2/graphene [14], MoS_2/CNTs [15], and MoS_2/C [16] have even more outstanding electrochemical properties than bare MS_2 nanoplates when they were used as anodes for LIBs. Incorporation of carbon materials improves the charge current, effective surface area, and chemical tolerance of MS_2 nanocrystals. MoS_2/graphene composite nanosheets were synthesized by a one-step solution phase method. The first charge and discharge capacities of the MoS_2/graphene composites were 2200 and 1300 m A h g^{-1}, respectively, which are higher than those of bare MoS_2 and graphene. In addition, the cycling behavior exhibited a reversible capacity of 1290 m A h g^{-1} for up to 50 cycles, while the reversible capacity of the bare MoS_2 electrode declined to 605 m A h g^{-1} after 50 cycles. The high capacity and enhanced cycling stability of the composite were attributed to effective and rapid charge carrier transport back and forth from the MoS_2 layers to

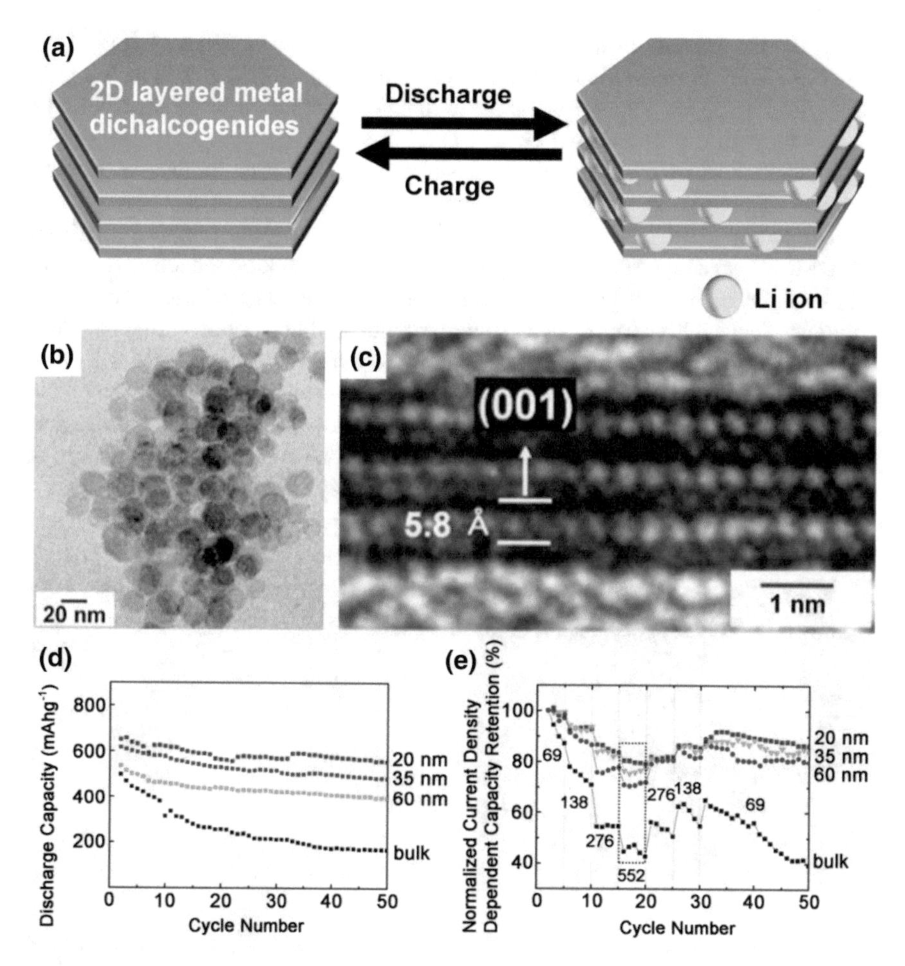

Fig. 3.1 **a** Schematic illustration of the reversible intercalation and exfoliation of Li^+ ions in 2-D layered anodic materials. **b** Top- and **c** side-view TEM images of as-synthesized ZrS_2 nanodiscs. **d** Cycling behaviors of ZrS_2 nanodiscs with different lateral dimensions (20, 35, and 60 nm) and bulk ZrS_2. **e** Current density-dependent capacity retention profiles of ZrS_2 nanodiscs (20, 35, and 60 nm) and bulk ZrS_2 at current densities of 69, 138, 276, and 552 mAh/g. Reproduced with permission from [9]. Copyright (2011) American Chemical Society

the graphene layer. Nanostructured SnS_2, although a promising anode material for LIBs, has the drawbacks of large volume changes and accompanying decreases in capacity during electrochemical cycling [17]. One solution to minimize the volume expansion is to distribute the tin-based materials evenly throughout another phase matrix. Zhi and co-workers prepared porous SnS_2/graphene composites through a two-step approach [12]. SnO_2 nanoparticles were formed on the surface of graphene nanosheets and then transformed into 2D SnS_2 nanoplates by reaction with H_2S gas.

The reversible capacity of the SnS_2/graphene composites was 650 m A h g^{-1} after 50 cycles, which is much higher than that of bare SnS_2 nanoplates (277 m A h g^{-1}). These results suggest that graphene layer functions both as a buffer matrix and a conducting pathway to improve cycling durability.

As another energy storage system, supercapacitors, also known as electrochemical capacitors, are considered as one of the most promising candidates due to its high power density, fast charge/discharge, very long cycle life, and relatively low cost. 2D layered structures comprising of MC and other conducting materials (i.e., MC/graphene) have been suggested to be an efficient and promising electrode materials for high performance supercapacitor. Many reports include VS_2 nanosheet, MoS_2, CoS_2 ellipsoid, CoS_2/graphene hybrid, NiS, and WS_2 nanosheet [10, 18–23]. For example, the availability of MC as a supercapacitor is provided in Fig. 3.2. The WS_2/rGO hybrids with layered structures proved its electrochemical performance of up to 350 F/g at a scan rate of 2 mV/s with an energy density of ~49 Wh/kg. In addition, the WS_2/rGO electrode stays stable after 1000 cycles.

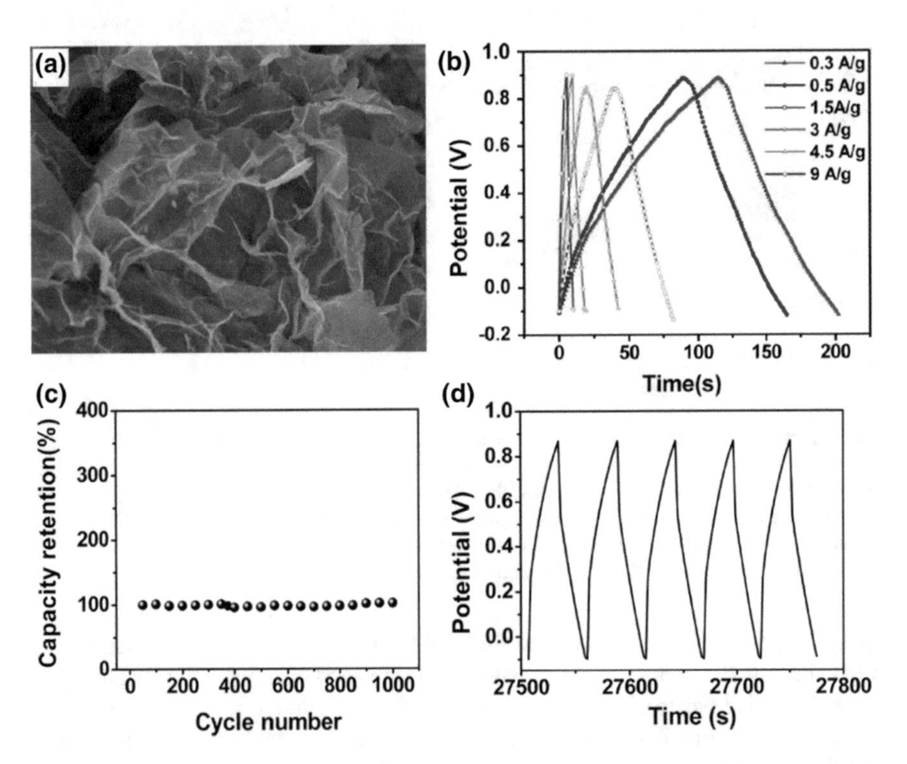

Fig. 3.2 **a** SEM image of WS_2 nanosheets. **b** Charge–discharge curves at different discharge current of WS_2/rGO hybrid electrode. **c** Cycling stability performance of the WS_2/rGO supercapacitor electrode at 3 A/g. **d** Last five charge-discharge cycles. Reproduced with permission from [10]. Copyright (2013) American Chemical Society

3.2 Thermoelectrics

Solid-state power generation and cooling systems based on thermoelectric effects have received great attention [24–27]. Thermoelectric devices composed of *p*- and *n*-type semiconductors can directly convert waste heat to electricity and vice versa. Bi_2Te_3 and its alloys with Se or Sb are some of the best thermoelectric materials at low temperatures ($\sim$80 °C) and PbTe-related materials are also strong candidates at slightly higher temperatures ($\sim$300 °C). Therefore, they have been widely used in low temperature power generation and small-scale cooling of electronic devices. The energy conversion efficiency of a thermoelectric device is evaluated by the dimensionless thermoelectric figure of merit (ZT), ZT = $S^2\sigma T/\kappa$, where S, σ, T, and κ are the Seebeck coefficient, electrical conductivity, temperature, and thermal conductivity, respectively. Recent theoretical and experimental advances have confirmed that introduction of nanoscale constituents increases the power factor ($S^2\sigma$) through a quantum confinement effect. The nano-sized grains reduce thermal conductivity (κ) more effectively than the reduction in electrical conductivity (σ). Furthermore, introduction of heterostructures in the nanocomposite can increase the Seebeck coefficient (S) due to a carrier filtering effect [27–30]. In this respect, better control over the grain size and shape of thermoelectric nanocrystals can be obtained using a bottom-up solution approach [31–36].

Wu and coworkers synthesized 1D barbell-shaped Bi_2Te_3–Te–Bi_2Te_3 heterostructures consisting of a Te nanowire and Bi_2Te_3 plates set at the two ends of the nanowire (Fig. 3.3a, b) [37]. They used a two-step conversion process. Te nanowires were synthesized by reducing tellurium dioxide with hydrazine hydrate solution. Once the formation of Te nanowires was completed, bismuth nitrate solution was hot-injected into the Te nanowire suspension at 160 °C, which allowed the Te nanowires to be converted into the barbell-shaped Te–Bi_2Te_3 heterostructures. By keeping the concentration of Bi low, Bi deposited to the axial growth tips and suppressed the random deposition on the surfaces in the radial direction. The heterostructured composite was fabricated by hot pressing. ZT of the heterostructured composite was two orders higher than that of a pure Te nanowire composite (Fig. 3.3b). The improved thermoelectric properties are mainly due to the enhanced Seebeck coefficient (S) caused by the carrier filtering and the decreased thermal conductivity (κ) caused by phonon scattering.

In terms of the grain boundaries to enhance the phonon scattering, thin 2D nanostructures such as nanoplates or nanosheets are promising because their face-to-face packing can generate tremendous boundaries along the pathway of the phonons. Several groups have investigated 2D Bi_2Te_3-related structures in the past decade [28]. Hyeon and coworkers synthesized Bi_2Te_3 nanoplates with a thickness of $\sim$1 nm. They investigated the thermoelectric properties of bulk pellets prepared by spark plasma sintering (Fig. 3.3c, d). Both the electrical (σ) and thermal (κ) conductivities increased as the sintering temperature was raised due to enhanced densification and grain growth during sintering. A maximum ZT value of 0.62 was achieved at 400 K from the nanoplate bulk pellets sintered at 250 °C (Fig. 3.3d).

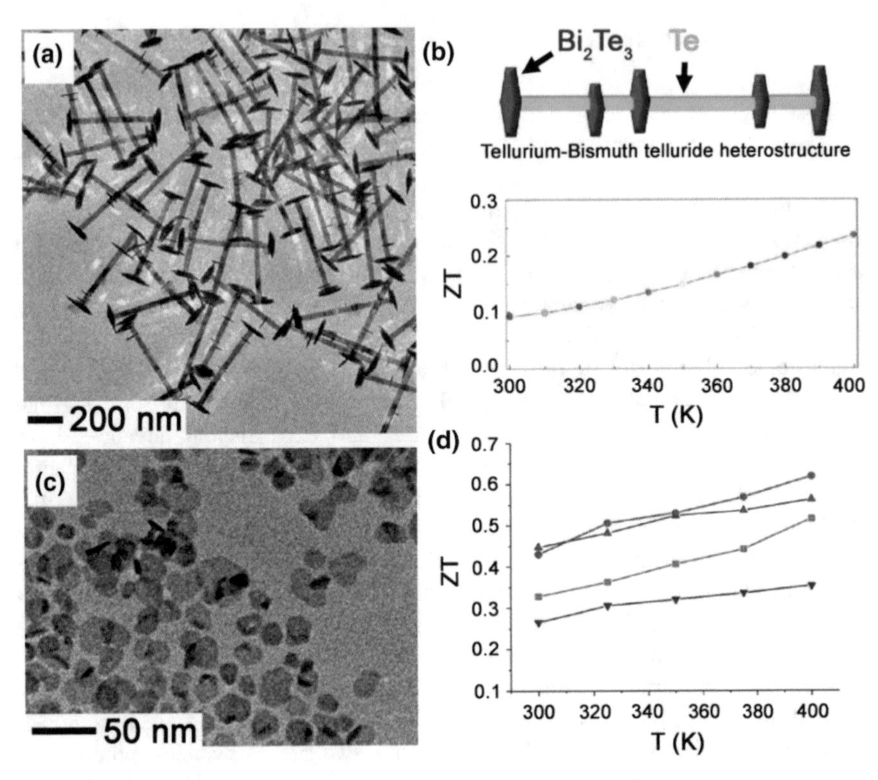

Fig. 3.3 Thermoelectric application of the Te–Bi_2Te_3 1-D heterostructure and Bi_2Te_3 2-D nanoplate. **a** TEM image of Te–Bi_2Te_3 heterostructure. **b** Schematic drawing of the Te–Bi_2Te_3 heterostructure and temperature dependence of the thermoelectric figure of merit (ZT) of the sintered bulk nanocomposite pellet. Reproduced with permission from [37]. Copyright (2012) American Chemical Society. **c** TEM image of the as-synthesized Bi_2Te_3 nanoplates. **d** Temperature dependence of the thermoelectric figure of merit of the Bi_2Te_3 nanoplates bulk pellets. The colors indicate the sintering temperature: 200 °C (green squares), 250 °C (red circles), 300 °C (blue upwards-pointing triangles), and 325 °C (purple downwards-pointing triangles). Reproduced with permission from [33]. Copyright (2012) American Chemical Society

Challenging issues associated with the use of nanocrystals for electric devices are removal of organic surfactant and thermal treatment without severely damaging the as-synthesized materials. A high sintering temperature (typically >400 °C) is required to decompose the organic surfactants. The insulating organic residue left behind by incomplete removal caused the material density of the pellet low and decreased the power factor of thermoelectric devices. Specifically, MC nanocrystals can form multi-component alloys at high sintering temperatures, which is not desirable for high-performance, reliable devices. Jeong and coworkers demonstrated scalable, high-yield production of surfactant-free Bi_2Te_3 and Bi_2Se_3 nanoflakes [28]. Simple mixing of the nanoflake suspensions allowed homogeneous distribution of the two nanocrystal species, hence fine control of the chemical composition of the Bi_2Te_3/Bi_2Se_3 nanocomposites was possible. The ZT value was 0.7 at 400 K in a broad range of Bi_2Se_3 contents (10–15 wt%), which is highly advantageous in preparing reliable devices because the ZT value of three component alloy nanocrystals ($Bi_2Te_{3-x}Se_x$) is too sensitive to the atomic composition of Se(x). The introduction of nanoscale grains and interfaces favors in terms of increase of phonon scattering and filtering low energy charge carriers without considerable reduction in electronic conductivity. As a proof-of-concept study, 2D heterostructured $Bi_2Se_3@Bi_2Te_3$ multishelled nanoplates was demonstrated in a solution-based epitaxial growth [38]. Fig. 3.4a shows the evolution of the seed Bi_2Se_3 nanoplates into core-shell, double-shell, and multishell bismuth chalcogenide nanoplates. Bismuth telluride is grown epitaxially on the Bi_2Se_3 seed nanoplates, predominantly on the side surfaces due to the intrinsic preference for 2D growth, with only a small fraction on the basal planes. The Moiré patterns in the core regions of the nanoplates (red dotted area) indicate that the cores overlap with a shell layer with different lattice distances or a crystal orientation (Fig. 3.4b).

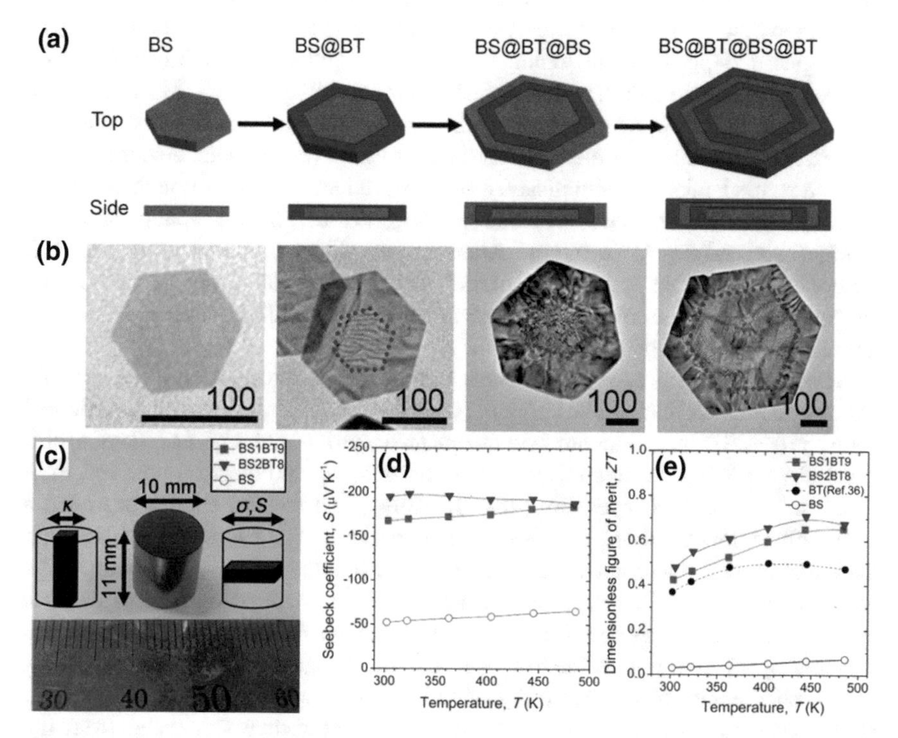

Fig. 3.4 **a** Evolution of the seed Bi_2Se_3 nanoplates into core-shell, double-shell, and multishell bismuth chalcogenide nanoplates $Bi_2Se_3@Bi_2Te_3@Bi_2Se_3@Bi_2Te_3$. **b** TEM images obtained from the nanoplates corresponding the epitaxial growth in (**a**). **c** A photograph of the sintered cylinder-shaped pellet sample. Temperature dependence of **c** Seebeck coefficient and **e** dimensionless thermoelectric figure of merit (ZT). Adapted with permission from [38]. Copyright (2015) American Chemical Society

The electrical conductivity of the sample (Fig. 3.4c) was measured in the range of 220–270 S/cm, which is relatively low due to the electron carrier scattering at the interfaces. The negative values of the Seebeck coefficient (S) indicates that electrons act as the major charge carriers (Fig. 3.4d). The κ value of the multishell nanoplates have lower, compared to thermal conductivity of Bi_2Te_3 bulk counterpart The simultaneous improvements in electronic and thermal transport properties of the multishell bismuth chalcogenide nanoplates obtained a high figure of merit (ZT) of $\sim$0.71 at 450 K (Fig. 3.4e).

3.3 Solar Cell

MC semiconductors have been considered promising materials for the absorber layer and charge transport layer in thin film photovoltaic devices such as CdTe solar cells and CI(G)Se solar cells. The chalcogenide films in the cells with a high power conversion efficiency (η) have been prepared by chemical bath deposition (CBD) or vapor phase depositions including evaporation, sputtering, sublimation, vapor transient deposition (VTD), and chemical vapor deposition (CVD) [39–41]. Recent advances in colloidal science have facilitated simple and eco-friendly synthetic approaches to produce various nanocrystals for use in photovoltaic devices. Use of the nanocrystal inks can significantly reduce production costs through the inexpensive deposition methods such as spraying, doctor blading, spin-casting, dip coating, and roll-to-roll printing. Although the films prepared from the MC nanocrystals has been considered inferior to the conventional films made by the vacuum processes, the solution-processed thin film solar cells have seen continuous improvement in the conversion efficiency.

Several research groups have devoted considerable effort toward synthesizing various colloidal nanocrystal inks [42–47]. In contrast to quasi-spherical nanocrystals, 1D nanostructures may enhance carrier mobility and device efficiency by reducing the frequency of electron hopping and electron-hole recombination in the absorber layer film [48–50]. Korgel and co-workers demonstrated a photovoltaic device made of $CuInSe_2$ nanowires [50]. The nanowires were 20 nm thick and several micrometers long. These were Indium-deficient with an average composition of $Cu_{1.0}In_{0.6}Se_{2.0}$. Free-standing fabric made of the nanowires was used to measure the photovoltaic response. Although the efficiency of the device was low (η = 0.1%) due to the low open circuit voltage (V_{oc}) and fill factor (FF), these results exhibited the feasibility of $CuInSe_2$ nanowires as the active layer for photovoltaic devices. The low power conversion efficiency originated from the voids and the random orientation of the nanowires, which reduced conductivity and prevented the carrier transport to their corresponding electrodes. The nanowire/quantum dot composite device enhanced transport connectivity. Conversion efficiency could potentially be improved by preparing defect-free nanowires and optimizing the alignment of the nanowires. Combining nanowires with spherical nanocrystals could also potentially improve device performance. Incorporating colloidal CdSe quantum dots into CdSe nanowires improved the conversion efficiency (η = 0.13%) photovoltaic performance [49]. A photovoltaic device comprising CdSe quantum dots and CdTe nanorods as the active layer has also been reported [51]. Due to the band alignment between the CdSe and CdTe nanocrystals, excited electrons transferred to the CdSe phase, while holes moved to the CdTe phase. Sintering of the nanorods enhanced the carrier mobility and the power conversion efficiency (η = 2.9%). Lin and coworkers utilized wurtzite-structured CdSe nanowires to fabricate a solar cell with the configuration of ITO/ZnO/CdSe NWs/PEDOT:PSS/Pt. They reported a conversion efficiency of 3.6% for this configuration [52].

Remarkable improvement of the power conversion efficiency has been achieved in the solar cells based on the I–III–VI_2 nanocrystals such as $CuInSe_2$ (CISe) and $Cu(In_xGa_{1-x})Se_2$ (CIGSe) [53–55]. Facile solution synthesis of CISe and related nanocrystal inks has been developed to prepare nanocrystal inks. In addition to 0-D nanocrystals, anisotropic 1D and 2D crystals in the forms of $CuInSe_2$ nanowires [56], $CuInSe_2$ nanoplatelets [57], $Cu_{2-x}S_ySe_{1-y}$ nanowires [58], $CuIn_{1-x}Ga_xS_2$ (CIGS) nanorods [59], and Cu_2ZnSnS_4 (CZTS) nanorods [60] have been synthesized by solvothermal method in the presence of the corresponding organometal precursors. Fine control of the chemical composition of the nanocrystals is the pre-requisite for a high performance device. Grain growth of the nanocrystals into high quality large crystals without incorporating micropores is another key factor. Agrawal and coworkers reported 5.5% efficiency using the CISe nanocrystal absorber layer after solid state densification of the layer during the selenization process [54]. Very recently, the same group prepared a CIGS absorber layer film and partially exchanged the S anions by Se to form CIGSSe layer. They reported 12% conversion efficiency by the process [53]. Jeong et al. achieved 8.2% efficiency with a CISe nanocrystal layer followed by high temperature densification [55]. In spite of the abrupt increase of the conversion efficiency, the chemical composition of ternary or quaternary MCs has yet to be controlled. And the effect of nanocrystal shape has not been studied. As it comes to the densification of the nanocrystal layer, in-situ self-assembly of the 1D or 2D nanocrystals into superstructures during the coating process may facilitate the formation of a non-porous crystal layer. The organic surfactants obstruct densification of the nanocrystals and the vapors cause elemental contamination which is not controlled. The synthesis of surfactant-free multicomponent nanocrystals is in great need [61].

Conjugated polymer-MC bulk heterojunction hybrid solar cells have been actively prepared during the last ten years [62–64]. These hybrid systems take advantage of the flexible nature of polymers and the wider light absorption of inorganic materials than organic materials. Most polymer-based hybrid solar cells consist of an interpenetrated network of electron donor and acceptor phases with bulk heterojunction interfaces. Alivisatos and co-workers first demonstrated a CdSe nanorod/poly(3-hexylthiophene) (P3HT) hybrid solar cell created by spin casting a solution of CdSe nanorods in P3HT [65]. Because CdSe and P3HT have complementary absorption spectra in the visible spectrum, devices composed of blended nanorod-polymer materials have a wide photocurrent spectrum from UV to 720 nm.

A power conversion efficiency of 1.7% was obtained. Since then, conversion efficiency has been steadily increased by adding additives to enhance the crystallinity of P3HT [66] or using different conjugated polymers [67, 68]. Conjugated polymers used widely in hybrid solar cells have relatively large band gaps ($\sim$1.9 eV), which limited absorption of the low energy solar spectrum from red to NIR. Low band gap polymers are promising candidates as an organic component for efficient light harvesting. Recently, CdSe tetrapods were combined with a low bandgap polymer (PCPDTBT), resulting in a power conversion efficiency of 3.13% under AM 1.5 illumination (Fig. 3.5) [69].

MC nanocrystals have also been used as light-absorbing sensitizers in organic-TiO_2 heterojunction hybrid solar cells. Although the performance of these solid-state DSSCs is lower than that of liquid-electrolyte-based DSSCs, the recent progress reported by Gratzel and coworkers is promising. This group generated Sb_2S_3 nanoparticles on a porous TiO_2 layer by CBD, and then the pores were filled with P3HT by spin-coating [70]. The resultant hybrid cell showed a high conversion efficiency (η = 5.13%) and highly stable photovoltaic performance in air without sealing the cell. Seok and coworkers further investigated the same system to find the best conducting polymer for hole injection from the Sb_2S_3 sensitizer nanocrystals. They achieved η = 6.18% at 1.5 AM 1.5 G radiation [71]. So far, spherical quantum dot sensitizers have been used attached to the surfaces of the porous metal oxide network. Anisotropic MCs such as nanorods and nanoplates deserve thorough investigation for the hole injection and chemical correlation with the polymer layer.

3.4 Resistive Switching

As one of non-volatile memories, resistive switching has been on the topic of future memory device since the first report on oxide insulators by Hickmott [72]. Typically, resistive switching is classified into thermal, electrical, or ion-migration-induced operation mechanism [73]. The ion-migration-induced

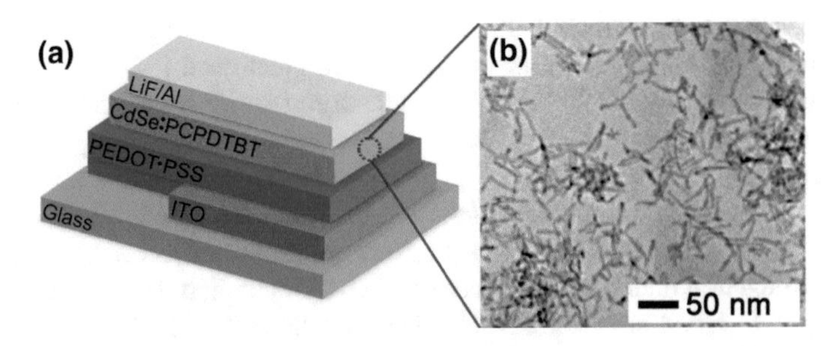

Fig. 3.5 **a** Photovoltaic device configuration of CdSe-PCPDTBT hybrid structure. **b** TEM image of CdSe tetrapod nanocrystals as the component of active layers. Reproduced with permission from [69]. Copyright (2010) American Chemical Society

change in resistance can occur by the migration of cation or anion in metal oxides and metal chalcogenides. Hirose et al., firstly, reported the switching behavior of Ag-photodoped As_2S_3 due to silver dendrite formation and annihilation [74]. Until recently, chalcogenide-based resistive switching such as GeSe:Ag [75], GeS:Cu [76], Ag_2S [77], $Ge_2Sb_3Te_5$ [78] was reported, which suggested the conductive filament growth through the channel as the mechanism for On/Off state development. Instead of thin films, anisotropic 1D metal chalcogenides can also be candidates for realization of resistive memory device; Ag_2S, Ag_2Se, Ag_2Te, Cu_2S, Cu_2Se, $Cu_{2-x}Se$ [79–84]. The high chemical reactivity of ultrathin Te nanowire provides a simple and robust way to fabricate ultrathin film of well-arrayed metal telluride with a tenability of stoichiometry [85]. Experimentally, the assembled Te NW films were transferred, repeatedly if needed, to a substrate. After drying, the Te NW films were immersed in $AgNO_3$ solution (ethylene glycol) to transform into Ag_2Te NW film. These Ag_2Te NW films were pressed by roller to make sure that the nanowires were in full contact with Au electrodes. Elemental Te is a well-known narrow band gap semiconductor (0.33 eV), which can be thought of as one of metalloids. Thus, current–voltage curve of Te NW film behaves like an ohmic material and experiences the transition to non-linear I–V curve formation during the incorporation of silver in the structure (Fig. 3.6a) [84]. Fig. 3.6b , c show the electrical properties by measuring current–voltage hysteresis of Ag_2Te NW films (monolayer and 5 layers of Ag_2TeNW films). The arrows in the figures represent the direction of the applied voltage scan. All the I–V curves from Ag_2Te NW films are non-linear and symmetric, indicating that the probe current of low resistance state is higher than the current of high resistance at the same applied voltage. As the number of Ag_2Te NW layers increases, the measured current of the device increases. The currents of all the Ag_2Te NW film devices suddenly jumped to high value and after that they slowly decreased or added small increment whenever the applied voltage reversed its polarity. Silver-deficient silver telluride (Ag_xTe, $x < 2$) can also be prepared, but still requires more silver ions in the crystal structure to bridge the current path for low resistance state (Fig. 3.6d). The calculated stoichiometries of the silver telluride NW films are $Ag_{0.08\pm0.06}Te$, $Ag_{0.24\pm0.12}Te$, $Ag_{1.67\pm0.16}Te$, and $Ag_{1.99\pm0.17}Te$, respectively. Electrical properties changed dramatically above the certain point (sample of $Ag_{0.24\pm0.12}Te$) and lost the initial ohmic characteristic (Fig. 3.6d).

As Aono group suggested, the electrically-induced conductive filament can be formed through metal cation migration and reduction in the case of silver sulfide (Ag_2S). In the same manner, the silver cation is enough to create the more conductive filament from $Ag_{X2}Te$ and the Ag_2Te reached its maximum value. the migration of silver cation and conductive formation was further investigated by testing non-stoichiometric $Ag_{2+\delta}Te$NW ($Ag_{2.03\pm0.19}Te$, $Ag_{2.06\pm0.21}Te$, $Ag_{2.21\pm0.49}Te$). Figure 3.6f shows the cycle test of On and OFF states of 5 layers of non-stoichiometric silver telluride NW film, which has no overlap between the high resistance and low resistance state. The On/Off ratio increased up to ~ 5 in the case of $Ag_{2.21\pm0.49}Te$NW sample. The physical origin of the resistive memory effect is suggested by the dynamics of electrons and ions inside structures. It has been

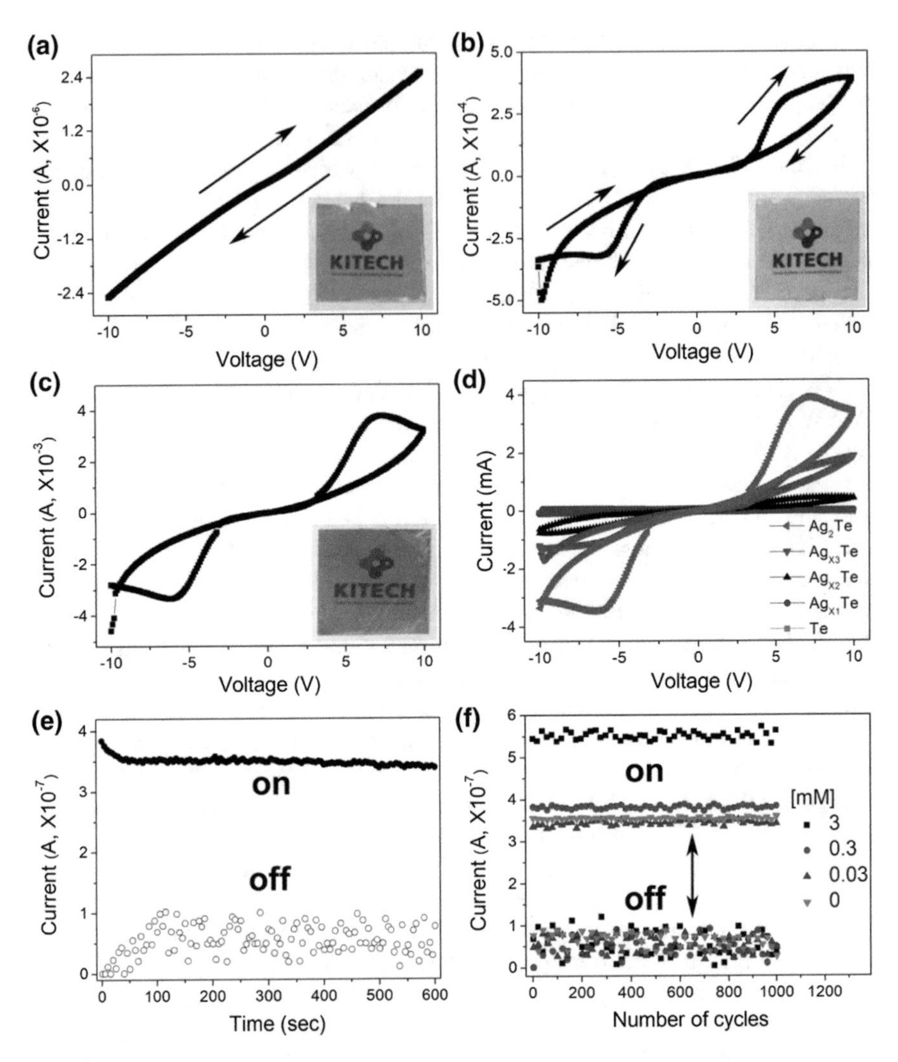

Fig. 3.6 Electrical properties of the Te NW and Ag_2Te NW film transferred to thermally-deposited gold electrodes. **a–d** I–V characteristics of NW films consisting of monolayer of Te NW (**a**), monolayer of Ag_2Te NW (**b**), five layers of Ag_2Te NW (**c**), respectively. **d** I–V characteristics of the silver telluride nanowire films consisting of 5 layers with different stoichiometries from Te, $Ag_{0.08\pm0.06}Te$, $Ag_{0.24\pm0.12}Te$, $Ag_{1.67\pm0.16}Te$, and $Ag_{1.99\pm0.17}Te$. **e** Long time response of the On state (solid circle) and Off state (open circle) as probed by the read voltage of 10 mV. **f** The ON and OFF states of five layers of Ag_2Te NW film. The high and low current states were induced by applying pulsed voltage of −8 and +8 V. The device current was probed by applying 10 mV. Reproduced by permission of the Royal Society of Chemistry [85]

reported that the effect is quite relevant at the nanoscale segments, which is explained by the conducting path formation and the stability of the high current state under electric field. The reversible electric properties of the Ag_2Te NW film can be explained by the conducting Ag segments inside Ag_2Te NWs as a charge storage center or vacancy migration under an electric field. In both explanations, it can be inferred that charge carriers are stored or accumulated at the interface and they can induce the high current state when the polarity is reversed. It is widely known that very thin inorganic nanostructures can bend and fold themselves without creating disastrous cracks or breakage due to good stress release originating from extremely high surface area. Thus, ultrathin nanowires are one of candidates for making up of components in flexible electronics. Despite of the high flexibility and resistive switching properties of ultrathin TeNWs, the assembled Te NW film still needs flexible working electrodes and substrates to fulfill all the requirements for a flexible device. Well-attached TeNW films on such flexible substrates were transformed to Ag_2TeNW film before completing working electrode configuration. The conventional thermally-deposited Au electrodes can go through cracks leading to the failure of charge transport at even a small strain. To demonstrate a flexible Ag_2Te resistive memory device consisting of full flexible components, we adopted AuNS electrode due to its high flexibility, bendability, and stretchability. In addition, the AuNS can also make a monolayer assembly on water surface, followed by transfer to substrates. The synthesized Au nanosheets (thickness $\approx$20 nm, lateral dimension $\geq$ 10 μm) in aqueous solution were redisperse in 1-butanol to spread on water surface. The device was bended and then flattened at every twenty cycles during measurement. The current values showed no substantial change between the flat and bend states, which proves the availability of the AuNS–Ag_2TeNW–AuNS device for a nanoscale and flexible resistive memory device.

3.5 Photodetector

Photodetectors that contain nanocrystal solids are some of the simplest optoelectronic devices. The electrical conductivity of the nanocrystal solids changes under illumination due to the increase in density of mobile charge carriers [86]. Photodetectors require high sensitivity, selectivity, and stability. Many MCs has high absorption coefficient and good photostability, which is excellent characteristic of photodetecting materials.

Various MC nanowires have been tested for the purpose because of their tunable electronic structure and band gap as well as the enhanced conduction of charge carriers in the length direction [88–97]. Among various 1D MC nanocrystals, silver sulfide (Ag_2S) is known as an excellent optical sensing material with a superior chemical stability. Li and co-workers reported a simple route to generate single crystalline Ag_2S nanowires that involves addition of sulfur powder into octadecylamine solution containing an $Ag(NO)_3$ precursor at 120 °C [97]. The photoelectric properties of the individual Ag_2S nanowires were investigated under UV illumination. The current increased sharply from 2.3 to 594 pA upon UV illumination, but the current decreased to dark current within 1 s upon turn-off of the UV light. Golberg and co-workers demonstrated a fast response photodetector made of

Sb_2Se_3 nanowires synthesized by a hydrothermal process [94]. 2D MC nanocrystals have also intriguing photoelectric properties originated from the 2D confined structures. The photocurrent conversion at the specific photon energies makes them great candidate for wavelength-selective photodetectors [87, 98–101]. Park and coworkers have demonstrated a band-selective photodetector based on molybdenum chalcogenides on flexible substrates, targeted towards visible range detection. Figure 3.7a shows a photodetector device consisting of $MoSe_2$ nanosheets exfoliated with $PS–NH_2$. The photocurrent increases with increasing power, producing a maximum of I_{on}/I_{off} ratio of $\sim 10^5$ at 238 mW/cm^2 and 10 V (Fig. 3.7b). The photo-excited carriers in conduction band of $MoSe_2$ drift to Au electrode with quasi-Ohmic contact under bias voltage. The increased photocurrent is mainly determined by the photoelectric effect, which is band-to-band excitation without any reversible charge loss. Molybdenum chalcogenides have shown the mechanical flexibility fabricated on a conventional filter paper (Fig. 3.7e).

Compared to conventional one-component (inorganic or organic) photodetectors, organic-inorganic hybrid photodetectors have unique features conferred by combination of the low ionization potential of organic molecules and the high electron affinity of inorganic components. These hybrid devices provide good physical flexibility and tunable functionality. Hybrid photodetectors with different combinations of organic and inorganic components have been fabricated [102–104]. Hybrid photodetectors composed of conjugated polymer, poly (3-hexylthiophene) (P3HT), and CdSe nanowires were fabricated on a rigid SiO/Si substrate, a flexible PET substrate, and printing paper [105]. These devices showed an enhanced photocurrent and more rapid response and recovery times than photodetectors fabricated using only one component. This was attributed to the high hole-transport rate of the polymer P3HT, the high electrical conductivity of the CdSe nanowires, and the synergistic effect of absorption spectra in the visible range. The devices fabricated on flexible substrates exhibited good flexibility, folding strength, excellent wavelength-dependent electrical stability, and rapid response to high-frequency light signals.

3.6 Electrocatalyst

Most research on the cathodes of fuel cells has focused on Pt based-materials because of their excellent electrocatalytic performance. However, Pt based-materials show severe Pt-dependence [106], the kinetics of the oxygen reduction reaction (ORR) are slow [107], and there is CO deactivation [108]. Various MCs are considered promising Pt-free ORR electrocatalysts because of their low cost, high tolerance to methanol and halide ions, and simple solution-based synthetic approach [109].

Nanoporous hollow cobalt sulfide nanosheet demonstrated high electrochemical oxygen evolution reaction (OER) activity and stability [110]. Hollow Co_3S_4 nanosheets can be synthesized by chemical transformation from $Co(OH)_2$

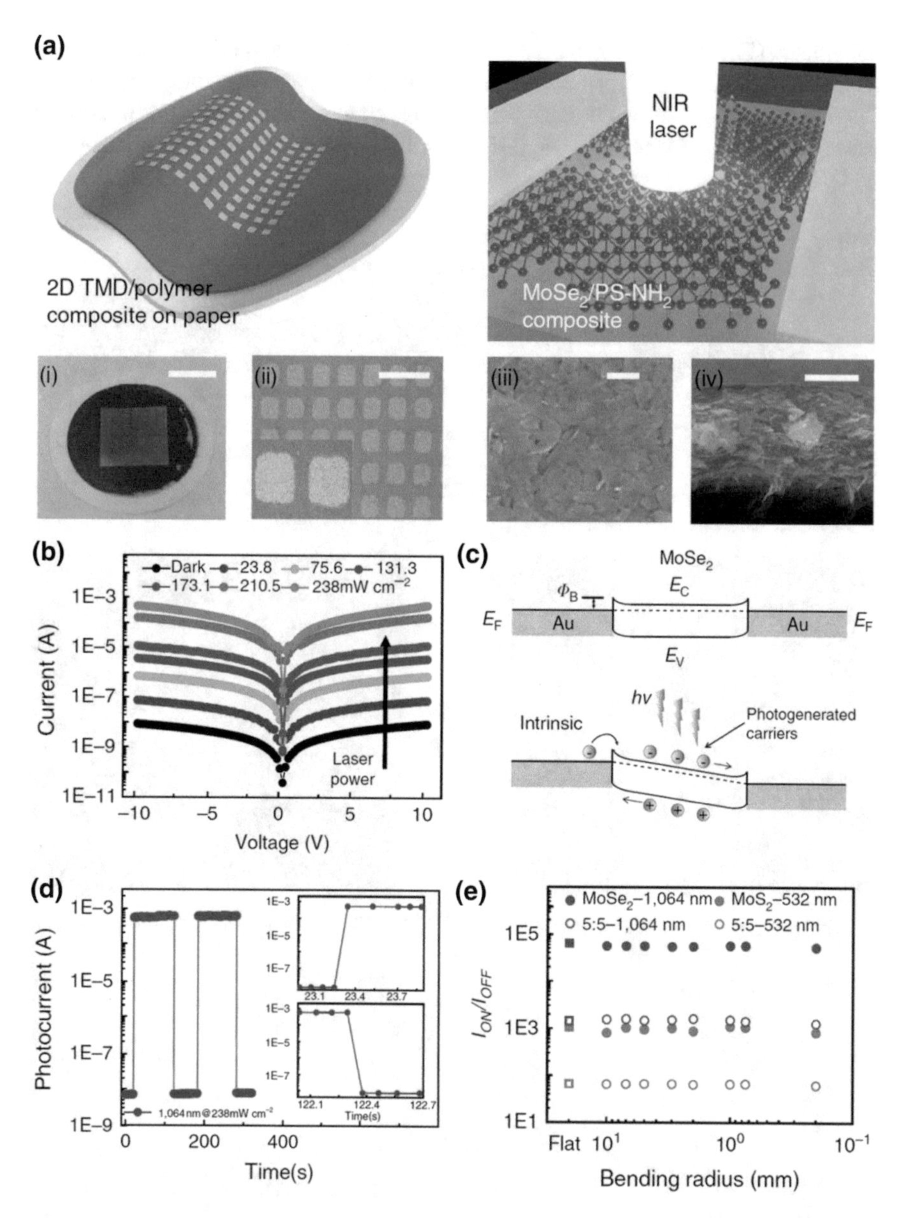

Fig. 3.7 **a** Schematic illustration and photographs of the two-terminal, parallel-type photodetector device consisting of $MoSe_2$ nanosheets exfoliated with PS–NH_2. SEM images of the surface and cross-section of the film stacked with each other. **b** Semi-log scale I–V characteristics of the $MoSe_2$ photodetector in the dark and under different light intensities of NIR at a wavelength of 1064 nm with a bias voltage of ±10 V. **c** Band diagram of the $MoSe_2$ photodetector. **d** Photoswitching behavior under alternating ON and OFF NIR light with an intensity of 238 mW/cm^2 with a switching time of ~100 ms. **e** Ratio of the photocurrent to the dark current of the $MoSe_2$, MoS_2, and $MoSe_2$/MoS_2 (5:5) composites as a function of the bending radius (1064 nm with 63 mW/cm^2 and 532 nm with 63 mW/cm^2). Reproduced with permission from [87]. Copyright (2015) Nature Communications

nanoplates (Fig. 3.8). The polarization curve shows the nanoporous hollow Co_3S_4 nanosheets produce greater current density and lower onset potential of catalytic current, compared to solid and commercial counterparts (Fig. 3.8d). Another example includes $CoSe_2$–DETA (DETA = diethylenetriamine) hybrid nanobelts

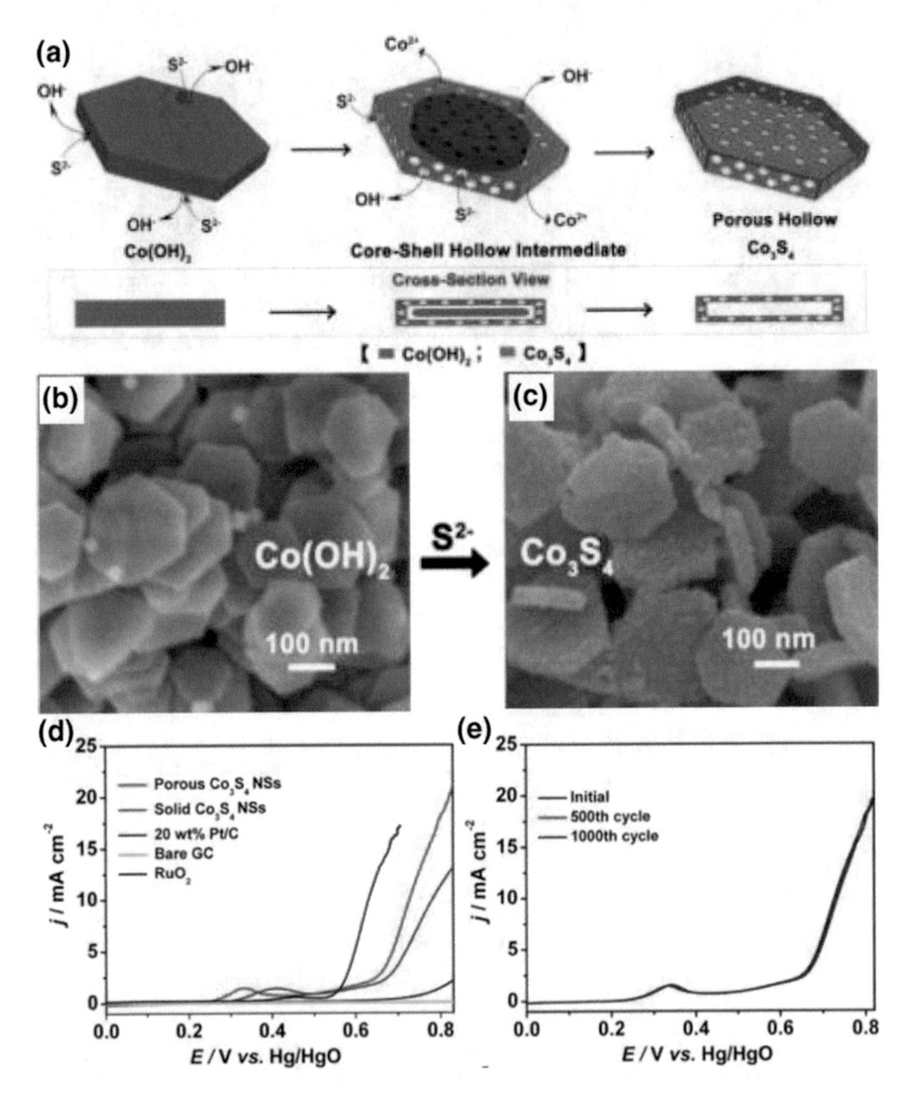

Fig. 3.8 Preparation of nanoporous Co_3S_4 nanosheets via anion exchange reaction. **a** Scheme of the chemical transformation from $Co(OH)_2$ to Co_3S_4. SEM images of **b** $Co(OH)_2$ and **c** Co_3S_4 nanosheets. **d** Polarization curves for OER on bare GC electrode and modified GC electrodes comprising the nanoporous Co_3S_4 nanosheets, solid Co_3S_4 nanosheets, Pt/C, and RuO_2, respectively. **e** OER polarization curves of nanoporous Co_3S_4 nanosheets before and after different cycles of accelerated stability test. Adapted with permission from [110]. Copyright (2014) American Chemical Society

that have shown good electrocatalytic performance with respect to ORR (onset potential of ca. 0.71 V) in acidic media. This good performance is ascribed to the large specific surface area and well-defined layered structure of $CoSe_2$–DETA constructed with small DETA molecules [111]. Long-term stability of the $CoSe_2$–DETA still needs to be addressed before these can be used practically. The properties of MC cathodes can be improved by introducing other nanoparticles. For example, adding Fe_3O_4 nanoparticles to the $CoSe_2$ hybrid nanobelt catalyst resulted in an increase of onset potential and current density. Although the electrocatalytic activity of this catalyst is still lower than that of Pt-based catalysts, these hybrid MC systems deserve further investigation.

3.7 Topological Insulator

Topological insulator is an unconventional quantum phase of semiconducting or insulating matter that possesses a metallic conductivity on their surfaces [112–114]. Surface electronic states are spin polarized and protected by time-reversal symmetry. Significant efforts have been devoted to investigate the potential application of topological insulators in spintronics and quantum information processing [115–117]. The number of publications on the layered MC materials, mainly V–VI (V = Bi, Sb; VI = Se, Te), has been increasing dramatically since these family of compounds were predicted to be three-dimensional topological insulators with unique surface states composed of a single Dirac cone at the Γ point [118, 119]. The strong spin-orbit coupling regulates robust and nontrivial surface states, which are topologically protected against back scattering from time-reversal invariant defects and impurities. In particular, the surface state of Bi_2Se_3 forms a single Dirac cone inside a large bulk band gap of 0.3 eV [119]. Fig. 3.9 shows a composite low magnification STM topography image with the topographic height of vapor-liquid-solid (VLS) grown Bi_2Se_3 nanoribbons. A line-cut height graph tells the height of two steps as 10.2 and 10.8 Å, respectively, indicating the Bi_2Se_3 quintuple layer thickness. High quality Bi_2Se_3 nanoribbons were adopted as topological insulator measurement by integrating them into a six terminal hall bar configuration (Fig. 3.9g) The resistance decreases with the temperature decrease and reaches a saturation minimum value below 20 K, which is consistent with a typical behavior of a heavily-doped semiconductor. The R_H-B curve with a small slope in Fig. 3.9h initially has linear dependence with the magnetic field up to 9 T, which corresponds to a high carrier area density. The R_H-B curve with a large slope in the low field region deviates from the linear behavior at high field, indicating the existence of different types of carriers [120]. The nonlinear dependence of R_H on the B field can be attributed to the presence of the surface states. A band calculation from recent reports shows that Ag_2Te can also be a topological insulator with an anisotropic single Dirac cone due to a distorted antifluorite structures [122].

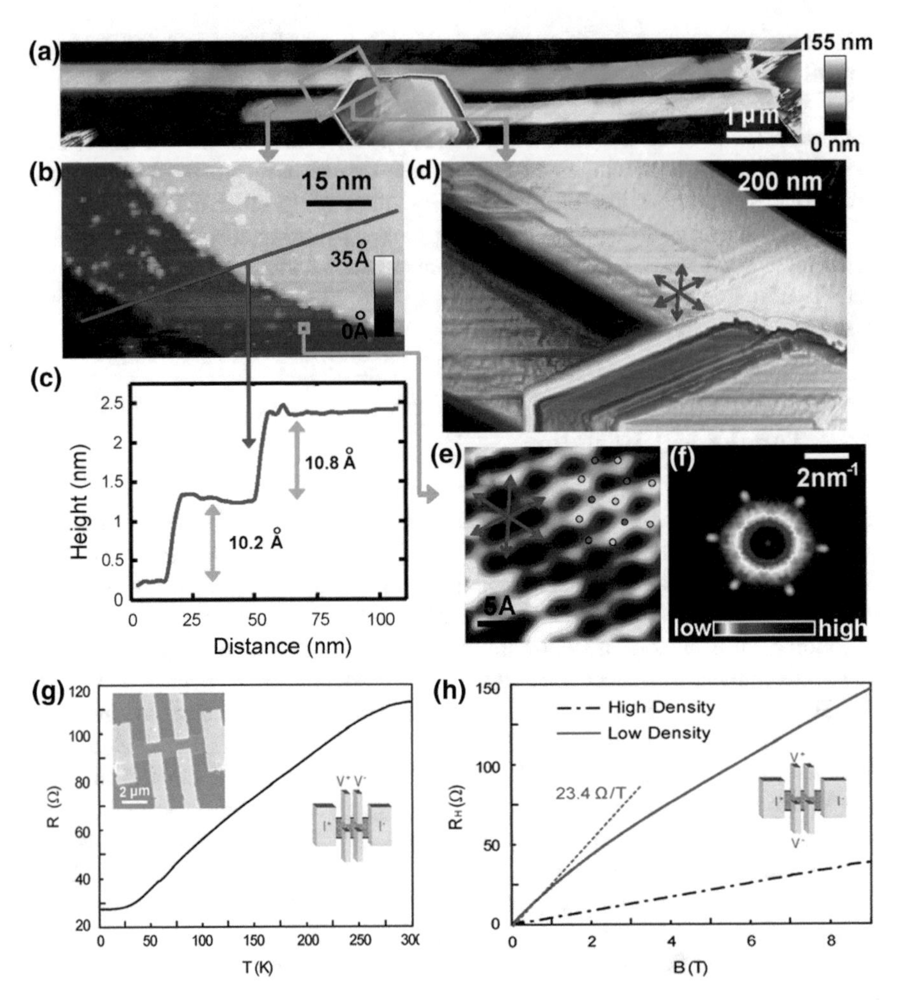

Fig. 3.9 **a** Composite STM topography of Bi_2Se_3 nanoribbons and a nanoplate. **b** Topography showing terraces on the nanoribbon surface. **c** Ribbon height across three steps [blue line in (**b**)]. **d** 3D topograph of nanoribbons and sheets. **e** Topography of honeycomb atomic lattice. **f** Magnitude of the Fourier transform of the atomically resolved topograph. **g** Temperature dependence of four-terminal resistance from RT to 2 K. The insets are SEM image of the Hall bar device and the measurement setting. **h** Hall traces measured at 2 K with the measurement setting in the inset. Reproduced with permission from [121]. Copyright (2010) American Chemical Society

Currently, most studies on topological insulators have been conducted with chalcogenides obtained through vapor-phase synthesis. Fine control of the thickness, defects, and atomic states at the surfaces are challenging issues. And the synthesis of wide 2D MC nanosheets (larger than at least a few micrometers) should be synthesized in a highly reproducible manner. And, the organic surfactants are serious obstacle of the solution-based MC nanocrystals in this area. Surfactant-free or inorganic surfactant systems during the preparation of MC nanocrystals create the clean surface and cause the negligible change of the surface electronic state. Solving these issues of synthesis is expected to draw advance in MC-based transistors.

3.8 Localized Surface Plasmon Resonance

Plasmonics has been increasing its impact on a various disciplines including quantum optics, photovoltaics, photocatalysts, metamaterials, and medicine. Surface plasmon is a light wave that evolves as a result of resonant interactions between the electromagnetic field of incident light and the plasma of electrons confined in a restricted dimension. Plasmons confined in nano-sized dimensions oscillate locally with a frequency known as the LSPR. The intensity and frequency of surface plasmon absorption bands are highly dependent on the characteristics of the material, such as the material species, size, shape, and additives on the surface [123]. Resonant waves are sensitive to the dielectric properties of the surrounding and can be tuned by external electromagnetic waves [124], which enables the surface plasmon to be utilized for sensing [125], imaging [126], information processing [127], and optoelectronic purposes [128]. LSPRs have been extensively characterized with metal nanoparticles (Au, Ag, Pt, Cu etc.) due to their enormous free carrier densities and sharp LSPR modes.

Other than metal nanocrystals, LSPRs have been reported from nonmetallic nanostructures including metal oxides and metal chalcogenides [101, 129, 131]. Such nonmetallic LSPR materials have an additional control variable in LSPR, which is tunability of the carrier concentration by doping. Doping plasmonic materials greatly expands the accessible absorbance range from visible to near infrared (NIR), even to mid-infrared regions (Fig. 3.10) [129]. The wider tunability of absorbance peaks can make it possible to introduce new devices such as electrochromic smart window, window coatings [132, 133]. Copper chalcogenide nanocrystals has been the species studied the most so far [134–137]. When Cu is slightly deficient with respect to S or Se, the nanocrystals contain a large number of free hole carriers in the valence band. The LSPR bands of CuS nanoparticles blue-shifted as the dopant concentration was increased. Recently, the effects of morphology on the LSPR response has been investigated in a $Cu_{2-x}S$ nanodisk system [130]. In-plane and out-of-plane dipolar resonances were determined. The wavelengths, line shapes, and relative intensities of theses plasmon bands were tuned by controlling the geometric aspect ratio of the disk or free carrier densities.

Figure 3.11a shows the NIR extinction spectra of $Cu_{2-x}S$ nanosphere and nanodisk by changing aspect ratio. The NIR spectrum of the nanodisk is characterized by two extinction bands. The multiple extinction peaks explains the excitation of shape-dependent dipolar LSPR modes, where excitation occurs in two discrete directions: parallel and perpendicular to the basal plane. In contrast, the nanosphere exhibits a single dipolar LSPR band. As the aspect ratio of the nanodisk increases,

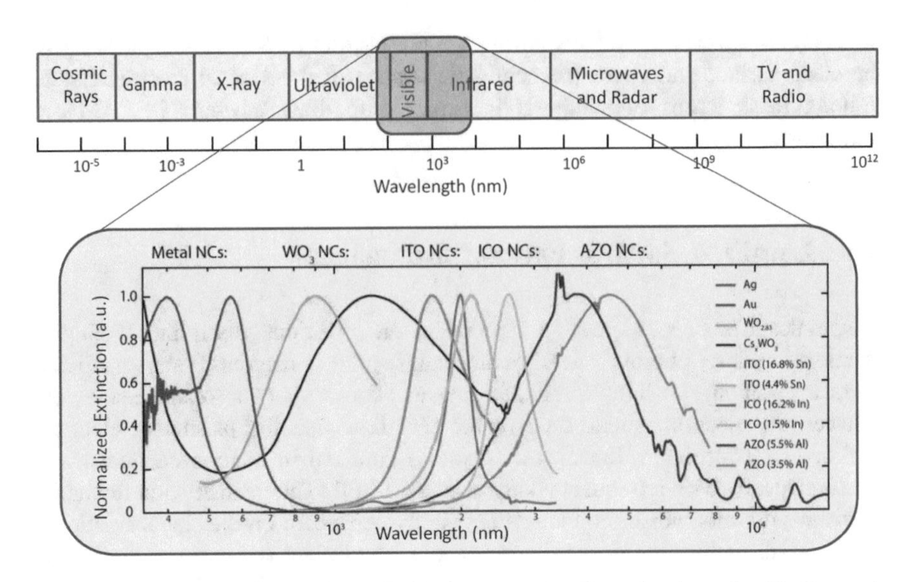

Fig. 3.10 Electromagnetic spectrum and absorbance spectra of metal and metal oxide plasmonic materials. All materials are spherical in shape with the exception of $WO_{2.83}$ rods. Reproduced with permission from [129]. Copyright (2014) John Wiley & Sons, Inc.

the LSPR wavelengths blue-shift to shorter wavelengths (Fig. 3.11b). It is contradictory to the experimental results that noble metals such as Au or Ag exhibit the red-shift with increase of aspect ratio. The observed blue-shift in the LSPR peaks during the increase of aspect ratio can be attributed to the following effects: an increase in nanodisk size and carrier density due to thermal oxidation. The calculated shift in the in-plane LSPR wavelength explains the competition behavior between the change of ω_{sp} from the Cu vacancies and the variation of ω_{sp} from aspect ratio (Fig. 3.11d). Furthermore, a rapid increase in the peak intensity associated with the out-of-plane LSPR mode is attributed to the increase of total surface area of the basal plane. In semiconductor LSPR materials, both shape effects and the free carrier density should be considered in controlling the LSPR properties.

3.9 Superconductor

Some MCs have recently emerged as a new class of superconductors that are electrically conductive without resistance below a certain temperature [138–140]. Iron chalcogenides (FeSe, FeTe) with a layered structure are considered high-superconducting transition temperature (T_c) superconductors [141]. In addition, unconventional superconducting states are expected from the interaction between magnetism and superconductivity (Fig. 3.12).

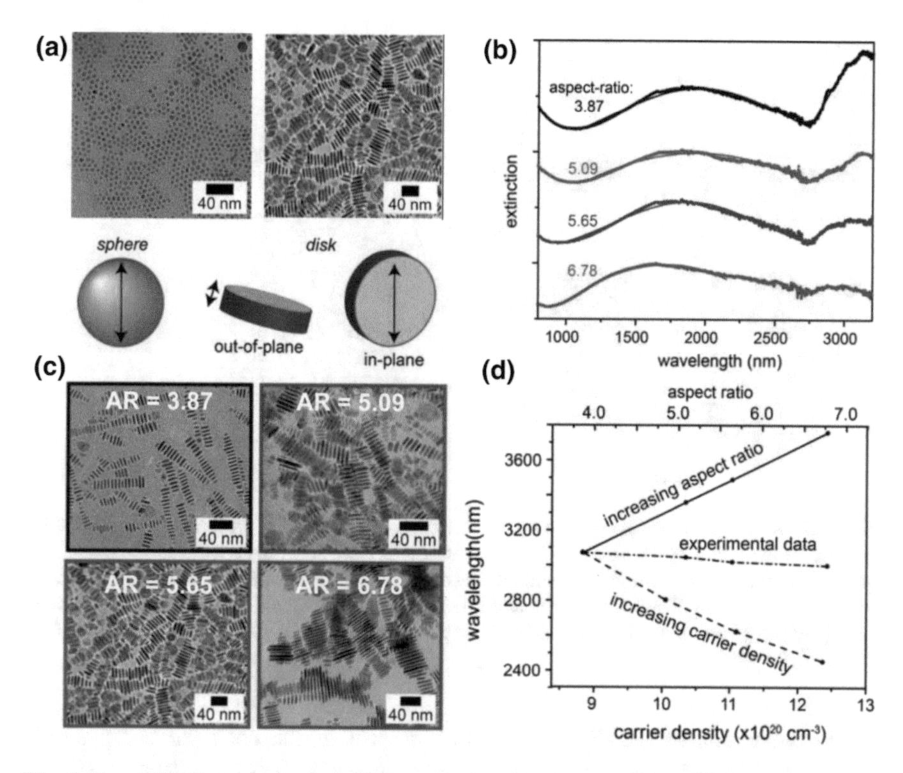

Fig. 3.11 **a** TEM images of spherical $Cu_{2-x}S$ and nanodisk. Scheme of LSPR polarization for spherical and disk-shaped nanocrystals. **b** Change of LSPR in $Cu_{2-x}S$ nanodisks with varying disk aspect ratio. **c** TEM images of the nanodisks with different aspect ratio. **d** Calculated change in LSPR wavelength for the in-plane mode for nanodisks with increasing aspect ratios (solid) and carrier densities (dashed), compared with the experimental data (dot-dash). Reproduced with permission [130]. Copyright (2011) American Chemical Society

Iron selenide (FeSe) has shown superconducting properties with a T_c of 8 K [142]. The T_c of FeSe has been improved up to $\sim$ 14 K by doping Te and to over 30 K under high pressure with a $\mathrm{d}T_c/\mathrm{d}P$ rate of $\sim$9.1 K Gpa^{-1} [143, 144]. Fig. 3.11 shows the temperature dependence of the resistance at different pressures in the α-$Fe_{1.93}Se_{0.57}Te_{0.43}$ system. Sulfur-doped $FeSe_{1-x}S_x$ systems ($x = 0.2$) have a T_c of 15.5 K [145]. Superconductivity at above 30 K was recently reported for $K_{0.8}Fe_2Se_2$ due to alkali intercalation between the FeSe layers [146]. Despite recent advances in iron based-superconductors, their chemical complexity, local structure, and the relationship between magnetism and superconductivity need to be understood to clarify the mechanism [147]. Unfortunately, no MCs synthesized by a solution phase approach have been reported to possess superconductivity.

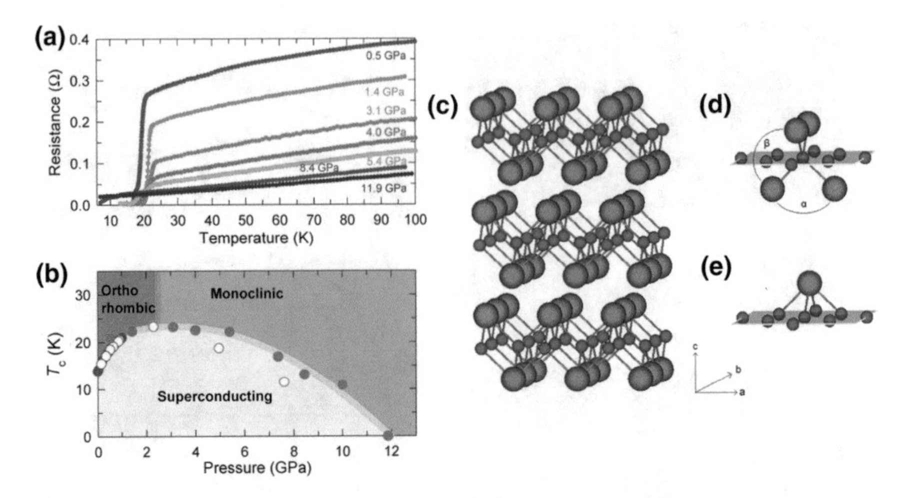

Fig. 3.12 **a** Temperature dependence of the resistance, *R*, at selected pressures up to 11.9 GPa. **b** Pressure evolution of T_c obtained from the magnetization and resistivity measurements as a function of applied pressure. **c** Schematic diagram of the crystal structure of α-$Fe_{1.93}Se_{0.57}Te_{0.43}$. Fe and Se/Te ions are displayed as blue and brown. Geometry of the **d** $Fe(Se/Te)_4$ tetrahedra and **e** the $(SE/Te)Fe_4$ pyramids. Reproduced with permission from [140]. Copyright (2009) American Chemical Society

We attribute this to the difficulty in fine control of the chemical composition, defects, and surface states in solution-based approaches. It is a challenging subject to realize the superconductivity in the materials synthesized in a solution.

References

1. Du G, Guo Z, Wang S, Zeng R, Chen Z, Liu H (2010) Superior stability and high capacity of restacked molybdenum disulfide as anode material for lithium ion batteries. Chem Commun 46(7):1106–1108. https://doi.org/10.1039/B920277C
2. Liu H, Su D, Wang G, Qiao SZ (2012) An ordered mesoporous WS2 anode material with superior electrochemical performance for lithium ion batteries. J Mater Chem 22(34):17437–17440. https://doi.org/10.1039/C2JM33992G
3. Seo JW, Jang JT, Park SW, Kim C, Park B, Cheon J (2008) Two-dimensional SnS_2 nanoplates with extraordinary high discharge capacity for lithium ion batteries. Adv Mater 20(22):4269–4273. https://doi.org/10.1002/adma.200703122
4. Hwang H, Kim H, Cho J (2011) MoS_2 nanoplates consisting of disordered graphene-like layers for high rate lithium battery anode materials. Nano Lett 11(11):4826–4830. https://doi.org/10.1021/nl202675f
5. Chang K, Chen W (2011) In situ synthesis of MoS_2/graphene nanosheet composites with extraordinarily high electrochemical performance for lithium ion batteries. Chem Commun 47(14):4252–4254. https://doi.org/10.1039/C1CC10631G

6. Altavilla C, Sarno M, Ciambelli P (2011) A novel wet chemistry approach for the synthesis of hybrid 2D free-floating single or multilayer nanosheets of MS_2@oleylamine (M=Mo, W). Chem Mater 23(17):3879–3885. https://doi.org/10.1021/cm200837g
7. Coleman JN, Lotya M, O'Neill A, Bergin SD, King PJ, Khan U, Young K, Gaucher A, De S, Smith RJ, Shvets IV, Arora SK, Stanton G, Kim H-Y, Lee K, Kim GT, Duesberg GS, Hallam T, Boland JJ, Wang JJ, Donegan JF, Grunlan JC, Moriarty G, Shmeliov A, Nicholls RJ, Perkins JM, Grieveson EM, Theuwissen K, McComb DW, Nellist PD, Nicolosi V (2011) Two-dimensional nanosheets produced by liquid exfoliation of layered materials. Science 331(6017):568–571. https://doi.org/10.1126/science.1194975
8. Cunningham G, Lotya M, Cucinotta CS, Sanvito S, Bergin SD, Menzel R, Shaffer MSP, Coleman JN (2012) Solvent exfoliation of transition metal dichalcogenides: dispersibility of exfoliated nanosheets varies only weakly between compounds. ACS Nano 6(4):3468–3480. https://doi.org/10.1021/nn300503e
9. Jang J-T, Jeong S, Seo J-W, Kim M-C, Sim E, Oh Y, Nam S, Park B, Cheon J (2011) Ultrathin zirconium disulfide nanodiscs. J Am Chem Soc 133(20):7636–7639. https://doi.org/10.1021/ja200400n
10. Ratha S, Rout CS (2013) Supercapacitor electrodes based on layered tungsten disulfide-reduced graphene oxide hybrids synthesized by a facile hydrothermal method. ACS Appl Mater Interfaces 5(21):11427–11433. https://doi.org/10.1021/am403663f
11. Xu C, Zeng Y, Rui X, Xiao N, Zhu J, Zhang W, Chen J, Liu W, Tan H, Hng HH, Yan Q (2012) Controlled soft-template synthesis of ultrathin C@FeS nanosheets with high-li-storage performance. ACS Nano 6(6):4713–4721. https://doi.org/10.1021/nn2045714
12. Luo B, Fang Y, Wang B, Zhou J, Song H, Zhi L (2012) Two dimensional graphene-SnS_2 hybrids with superior rate capability for lithium ion storage. Energy Environ Sci 5(1):5226–5230. https://doi.org/10.1039/C1EE02800F
13. Luo W, Xie X, Wu C, Zheng F (2008) Spherical CoS2@carbon core–shell nanoparticles: one-pot synthesis and Li storage property. Nanotechnology 19(7):075602. https://doi.org/10.1088/0957-4484/19/7/075602
14. Chang K, Chen W (2011) l-cysteine-assisted synthesis of layered MoS_2/graphene composites with excellent electrochemical performances for lithium ion batteries. ACS Nano 5(6):4720–4728. https://doi.org/10.1021/nn200659w
15. Ding S, Chen JS, Lou XW (2011) Glucose assisted growth of MoS_2 nanosheets on CNT backbone for improved lithium storage properties. Chem A Eur J 17 (47):13142–13145. https://doi.org/10.1002/chem.201102480
16. Das SK, Mallavajula R, Jayaprakash N, Archer LA (2012) Self-assembled MoS_2-carbon nanostructures: influence of nanostructuring and carbon on lithium battery performance. J Mater Chem 22(26):12988–12992. https://doi.org/10.1039/C2JM32468G
17. Idota Y, Kubota T, Matsufuji A, Maekawa Y, Miyasaka T (1997) Tin-based amorphous oxide: a high-capacity lithium-ion-storage material. Science 276(5317):1395–1397. https://doi.org/10.1126/science.276.5317.1395
18. Feng J, Sun X, Wu C, Peng L, Lin C, Hu S, Yang J, Xie Y (2011) Metallic few-layered vs2 ultrathin nanosheets: high two-dimensional conductivity for in-plane supercapacitors. J Am Chem Soc 133(44):17832–17838. https://doi.org/10.1021/ja207176c
19. Cao L, Yang S, Gao W, Liu Z, Gong Y, Ma L, Shi G, Lei S, Zhang Y, Zhang S, Vajtai R, Ajayan PM (2013) Direct laser-patterned micro-supercapacitors from paintable MoS_2 films. Small 9(17):2905–2910. https://doi.org/10.1002/smll.201203164
20. Zhang L, Wu HB, Lou XW (2012) Unusual CoS_2 ellipsoids with anisotropic tube-like cavities and their application in supercapacitors. Chem Commun 48(55):6912–6914. https://doi.org/10.1039/C2CC32750C

21. Wang Q, Jiao L, Du H, Yang J, Huan Q, Peng W, Si Y, Wang Y, Yuan H (2011) Facile synthesis and superior supercapacitor performances of three-dimensional cobalt sulfide hierarchitectures. CrystEngComm 13(23):6960–6963. https://doi.org/10.1039/C1CE06082A
22. Wang B, Park J, Su D, Wang C, Ahn H, Wang G (2012) Solvothermal synthesis of CoS_2-graphene nanocomposite material for high-performance supercapacitors. J Mater Chem 22 (31):15750–15756. https://doi.org/10.1039/C2JM31214J
23. Yang J, Duan X, Qin Q, Zheng W (2013) Solvothermal synthesis of hierarchical flower-like [small beta]-NiS with excellent electrochemical performance for supercapacitors. J Mater Chem A 1(27):7880–7884. https://doi.org/10.1039/C3TA11167A
24. Bell LE (2008) Cooling, heating, generating power, and recovering waste heat with thermoelectric systems. Science 321(5895):1457–1461. https://doi.org/10.1126/science.1158899
25. Sales BC (2002) Smaller is cooler. Science 295(5558):1248–1249. https://doi.org/10.1126/science.1069895
26. Dresselhaus MS, Chen G, Tang MY, Yang RG, Lee H, Wang DZ, Ren ZF, Fleurial JP, Gogna P (2007) New directions for low-dimensional thermoelectric materials. Adv Mater 19 (8):1043–1053. https://doi.org/10.1002/adma.200600527
27. Snyder GJ, Toberer ES (2008) Complex thermoelectric materials. Nat Mater 7:105. https://doi.org/10.1038/nmat2090
28. Min Y, Roh JW, Yang H, Park M, Kim SI, Hwang S, Lee SM, Lee KH, Jeong U (2013) Surfactant-free scalable synthesis of Bi_2Te_3 and Bi_2Se_3 nanoflakes and enhanced thermoelectric properties of their nanocomposites. Adv Mater 25(10):1425–1429. https://doi.org/10.1002/adma.201203764
29. Kim SI, Ahn K, Yeon D-H, Hwang S, Kim H-S, Lee SM, Lee KH (2011) Enhancement of seebeck coefficient in $Bi_{0.5}Sb_{1.5}Te_3$ with high-density tellurium nanoinclusions. Appl Phys Exp 4(9):091801. https://doi.org/10.1143/apex.4.091801
30. Ko D-K, Kang Y, Murray CB (2011) Enhanced thermopower via carrier energy filtering in solution-processable Pt–Sb_2Te_3 nanocomposites. Nano Lett 11(7):2841–2844. https://doi.org/10.1021/nl2012246
31. Scheele M, Oeschler N, Veremchuk I, Reinsberg K-G, Kreuziger A-M, Kornowski A, Broekaert J, Klinke C, Weller H (2010) ZT enhancement in solution-grown $Sb(2-x)BixTe_3$ nanoplatelets. ACS Nano 4(7):4283–4291. https://doi.org/10.1021/nn1008963
32. Soni A, Yanyuan Z, Ligen Y, Aik MKK, Dresselhaus MS, Xiong Q (2012) Enhanced thermoelectric properties of solution grown $Bi_2Te_{3-x}Sex$ nanoplatelet composites. Nano Lett 12(3):1203–1209. https://doi.org/10.1021/nl2034859
33. Son JS, Choi MK, Han M-K, Park K, Kim J-Y, Lim SJ, Oh M, Kuk Y, Park C, Kim S-J, Hyeon T (2012) n-type nanostructured thermoelectric materials prepared from chemically synthesized ultrathin Bi_2Te_3 nanoplates. Nano Lett 12(2):640–647. https://doi.org/10.1021/nl203389x
34. Mehta RJ, Zhang Y, Karthik C, Singh B, Siegel RW, Borca-Tasciuc T, Ramanath G (2012) A new class of doped nanobulk high-figure-of-merit thermoelectrics by scalable bottom-up assembly. Nat Mater 11:233. https://doi.org/10.1038/nmat3213
35. Zhang G, Kirk B, Jauregui LA, Yang H, Xu X, Chen YP, Wu Y (2012) Rational synthesis of ultrathin n-Type Bi_2Te_3 nanowires with enhanced thermoelectric properties. Nano Lett 12 (1):56–60. https://doi.org/10.1021/nl202935k
36. Liang W, Rabin O, Hochbaum AI, Fardy M, Zhang M, Yang P (2009) Thermoelectric properties of p-type PbSe nanowires. Nano Res 2(5):394–399. https://doi.org/10.1007/s12274-009-9039-2
37. Zhang G, Fang H, Yang H, Jauregui LA, Chen YP, Wu Y (2012) Design principle of telluride-based nanowire heterostructures for potential thermoelectric applications. Nano Lett 12(7):3627–3633. https://doi.org/10.1021/nl301327d

38. Min Y, Park G, Kim B, Giri A, Zeng J, Roh JW, Kim SI, Lee KH, Jeong U (2015) Synthesis of multishell nanoplates by consecutive epitaxial growth of Bi_2Se_3 and Bi_2Te_3 nanoplates and enhanced thermoelectric properties. ACS Nano 9(7):6843–6853. https://doi.org/10.1021/nn507250r
39. Wu X (2004) High-efficiency polycrystalline CdTe thin-film solar cells. Sol Energy 77 (6):803–814. https://doi.org/10.1016/j.solener.2004.06.006
40. Romeo A, Terheggen M, Abou-Ras D, Bätzner DL, Haug FJ, Kälin M, Rudmann D, Tiwari AN (2004) Development of thin-film Cu(In, Ga)Se_2 and CdTe solar cells. Prog Photovoltaics Res Appl 12(2–3):93–111. https://doi.org/10.1002/pip.527
41. Kemell M, Ritala M, Leskelä M (2005) Thin film deposition methods for CuInSe 2 solar cells. Crit Rev Solid State Mater Sci 30(1):1–31. https://doi.org/10.1080/10408430590918341
42. Jiang C, Lee J-S, Talapin DV (2012) Soluble Precursors for $CuInSe_2$, $CuIn_{1-x}GaxSe_2$, and $Cu_2ZnSn(S, Se)_4$ based on colloidal nanocrystals and molecular metal chalcogenide surface ligands. J Am Chem Soc 134(11):5010–5013. https://doi.org/10.1021/ja2105812
43. Steinhagen C, Panthani MG, Akhavan V, Goodfellow B, Koo B, Korgel BA (2009) Synthesis of Cu_2ZnSnS_4 nanocrystals for use in low-cost photovoltaics. J Am Chem Soc 131 (35):12554–12555. https://doi.org/10.1021/ja905922j
44. Guo Q, Hillhouse HW, Agrawal R (2009) Synthesis of Cu_2ZnSnS_4 nanocrystal ink and its use for solar cells. J Am Chem Soc 131(33):11672–11673. https://doi.org/10.1021/ja904981r
45. Weil BD, Connor ST, Cui Y (2010) $CuInS_2$ solar cells by air-stable ink rolling. J Am Chem Soc 132(19):6642–6643. https://doi.org/10.1021/ja1020475
46. Panthani MG, Akhavan V, Goodfellow B, Schmidtke JP, Dunn L, Dodabalapur A, Barbara PF, Korgel BA (2008) Synthesis of $CuInS_2$, $CuInSe_2$, and $Cu(InxGa1-x)Se_2$ (CIGS) nanocrystal "Inks" for printable photovoltaics. J Am Chem Soc 130(49):16770–16777. https://doi.org/10.1021/ja805845q
47. Puthussery J, Seefeld S, Berry N, Gibbs M, Law M (2011) Colloidal iron pyrite (FeS_2) nanocrystal inks for thin-film photovoltaics. J Am Chem Soc 133(4):716–719. https://doi.org/10.1021/ja1096368
48. Hochbaum AI, Yang P (2010) Semiconductor nanowires for energy conversion. Chem Rev 110(1):527–546. https://doi.org/10.1021/cr900075v
49. Yu Y, Kamat PV, Kuno M (2010) A CdSe nanowire/quantum dot hybrid architecture for improving solar cell performance. Adv Func Mater 20(9):1464–1472. https://doi.org/10.1002/adfm.200902372
50. Steinhagen C, Akhavan VA, Goodfellow BW, Panthani MG, Harris JT, Holmberg VC, Korgel BA (2011) Solution−liquid−solid synthesis of $CuInSe_2$ nanowires and their implementation in photovoltaic devices. ACS Appl Mater Interfaces 3(5):1781–1785. https://doi.org/10.1021/am200334d
51. Gur I, Fromer NA, Geier ML, Alivisatos AP (2005) Air-stable all-inorganic nanocrystal solar cells processed from solution. Science 310(5747):462–465. https://doi.org/10.1126/science.1117908
52. Feng Z, Zhang Q, Lin L, Guo H, Zhou J, Lin Z (2010) ⟨0001⟩-Preferential growth of CdSe nanowires on conducting glass: template-free electrodeposition and application in photovoltaics. Chem Mater 22(9):2705–2710. https://doi.org/10.1021/cm901703d
53. Guo Q, Ford GM, Agrawal R, Hillhouse HW (2013) Ink formulation and low-temperature incorporation of sodium to yield 12% efficient Cu(In, Ga)(S, Se)$_2$ solar cells from sulfide nanocrystal inks. Prog Photovoltaics Res Appl 21(1):64–71. https://doi.org/10.1002/pip.2200
54. Guo Q, Ford GM, Hillhouse HW, Agrawal R (2009) Sulfide nanocrystal inks for dense Cu (In1−xGax)(S1−ySey)$_2$ absorber films and their photovoltaic performance. Nano Lett 9 (8):3060–3065. https://doi.org/10.1021/nl901538w

55. Jeong S, Lee B-S, Ahn S, Yoon K, Seo Y-H, Choi Y, Ryu B-H (2012) An 8.2% efficient solution-processed $CuInSe_2$ solar cell based on multiphase $CuInSe_2$ nanoparticles. Energy Environ Sci 5(6):7539–7542. https://doi.org/10.1039/c2ee21269b
56. Wooten AJ, Werder DJ, Williams DJ, Casson JL, Hollingsworth JA (2009) Solution−liquid −solid growth of ternary Cu−In−Se semiconductor nanowires from multiple- and single-source precursors. J Am Chem Soc 131(44):16177–16188. https://doi.org/10.1021/ja905730n
57. Bi W, Zhou M, Ma Z, Zhang H, Yu J, Xie Y (2012) $CuInSe_2$ ultrathin nanoplatelets: novel self-sacrificial template-directed synthesis and application for flexible photodetectors. Chem Commun 48(73):9162–9164. https://doi.org/10.1039/C2CC34727J
58. Xu J, Tang YB, Chen X, Luan CY, Zhang WF, Zapien JA, Zhang WJ, Kwong HL, Meng XM, Lee ST, Lee CS (2010) Synthesis of homogeneously alloyed Cu2−x(SySe1−y) nanowire bundles with tunable compositions and bandgaps. Adv Func Mater 20(23):4190–4195. https://doi.org/10.1002/adfm.201000771
59. Singh A, Coughlan C, Laffir F, Ryan KM (2012) Assembly of $CuIn1_{-}xGaxS_2$ nanorods into highly ordered 2D and 3D superstructures. ACS Nano 6(8):6977–6983. https://doi.org/10.1021/nn301999b
60. Singh A, Geaney H, Laffir F, Ryan KM (2012) Colloidal synthesis of Wurtzite Cu_2ZnSnS_4 nanorods and their perpendicular assembly. J Am Chem Soc 134(6):2910–2913. https://doi.org/10.1021/ja2112146
61. Min Y, Moon GD, Park J, Park M, Jeong U (2011) Surfactant-free $CuInSe_2$ nanocrystals transformed from In 2 Se 3 nanoparticles and their application for a flexible UV photodetector. Nanotechnology 22(46):465604. https://doi.org/10.1088/0957-4484/22/46/465604
62. Xu T, Qiao Q (2011) Conjugated polymer-inorganic semiconductor hybrid solar cells. Energy Environ Sci 4(8):2700–2720. https://doi.org/10.1039/C0EE00632G
63. McGehee MD (2011) Nanostructured organic-inorganic hybrid solar cells. MRS Bull 34 (2):95–100. https://doi.org/10.1557/mrs2009.27
64. Fan X, Zhang M, Wang X, Yang F, Meng X (2013) Recent progress in organic-inorganic hybrid solar cells. J Mater Chem A 1(31):8694–8709. https://doi.org/10.1039/C3TA11200D
65. Huynh WU, Dittmer JJ, Alivisatos AP (2002) Hybrid nanorod-polymer solar cells. Science 295(5564):2425–2427. https://doi.org/10.1126/science.1069156
66. Wu Y, Zhang G (2010) Performance enhancement of hybrid solar cells through chemical vapor annealing. Nano Lett 10(5):1628–1631. https://doi.org/10.1021/nl904095n
67. Sun B, Snaith HJ, Dhoot AS, Westenhoff S, Greenham NC (2005) Vertically segregated hybrid blends for photovoltaic devices with improved efficiency. J Appl Phys 97(1):014914. https://doi.org/10.1063/1.1804613
68. Wang P, Abrusci A, Wong HMP, Svensson M, Andersson MR, Greenham NC (2006) Photoinduced charge transfer and efficient solar energy conversion in a blend of a red polyfluorene copolymer with CdSe nanoparticles. Nano Lett 6(8):1789–1793. https://doi.org/10.1021/nl061085q
69. Dayal S, Kopidakis N, Olson DC, Ginley DS, Rumbles G (2010) Photovoltaic devices with a low band gap polymer and CdSe nanostructures exceeding 3% efficiency. Nano Lett 10 (1):239–242. https://doi.org/10.1021/nl903406s
70. Chang JA, Rhee JH, Im SH, Lee YH, H-j Kim, Seok SI, Nazeeruddin MK, Gratzel M (2010) High-performance nanostructured inorganic−organic heterojunction solar cells. Nano Lett 10(7):2609–2612. https://doi.org/10.1021/nl101322h
71. Im SH, Lim C-S, Chang JA, Lee YH, Maiti N, Kim H-J, Nazeeruddin MK, Grätzel M, Seok SI (2011) Toward interaction of sensitizer and functional moieties in hole-transporting materials for efficient semiconductor-sensitized solar cells. Nano Lett 11(11):4789–4793. https://doi.org/10.1021/nl2026184
72. Hickmott TW (1962) Low-frequency negative resistance in thin anodic oxide films. J Appl Phys 33(9):2669–2682. https://doi.org/10.1063/1.1702530

73. Waser R, Aono M (2007) Nanoionics-based resistive switching memories. Nat Mater 6:833. https://doi.org/10.1038/nmat2023
74. Hirose Y, Hirose H (1976) Polarity-dependent memory switching and behavior of Ag dendrite in Ag-photodoped amorphous As_2S_3 films. J Appl Phys 47(6):2767–2772. https://doi.org/10.1063/1.322942
75. Kozicki MN, Mitkova M (2006) Mass transport in chalcogenide electrolyte films—materials and applications. J Non-Cryst Solids 352(6):567–577. https://doi.org/10.1016/j.jnoncrysol.2005.11.065
76. Fujii T, Arita M, Takahashi Y, Fujiwara I (2011) In situ transmission electron microscopy analysis of conductive filament during solid electrolyte resistance switching. Appl Phys Lett 98(21):212104. https://doi.org/10.1063/1.3593494
77. Xu Z, Bando Y, Wang W, Bai X, Golberg D (2010) Real-time in situ HRTEM-resolved resistance switching of Ag_2S nanoscale ionic conductor. ACS Nano 4(5):2515–2522. https://doi.org/10.1021/nn100483a
78. Pradel A, Frolet N, Ramonda M, Piarristeguy A, Ribes M (2011) Bipolar resistance switching in chalcogenide materials. Phys Status Solidi (a) 208(10):2303–2308. https://doi.org/10.1002/pssa.201000767
79. Lee NJ, An BH, Koo AY, Ji HM, Cho JW, Choi YJ, Kim YK, Kang CJ (2011) Resistive switching behavior in a Ni–Ag_2Se–Ni nanowire. Appl Phys A 102(4):897–900. https://doi.org/10.1007/s00339-011-6319-y
80. Schoen DT, Xie C, Cui Y (2007) Electrical switching and phase transformation in silver selenide nanowires. J Am Chem Soc 129(14):4116–4117. https://doi.org/10.1021/ja068365s
81. Rehman S, Kim K, Hur J-H, D-k Kim (2017) Phase transformation induced resistive switching behavior in Al/Cu 2 Se/Pt. J Phys D Appl Phys 50(13):135301. https://doi.org/10.1088/1361-6463/aa593e
82. Wu C-Y, Wu Y-L, Wang W-J, Mao D, Yu Y-Q, Wang L, Xu J, Hu J-G, Luo L-B (2013) High performance nonvolatile memory devices based on $Cu_2{-}x$Se nanowires. Appl Phys Lett 103(19):193501. https://doi.org/10.1063/1.4828881
83. Liu X, Mayer MT, Wang D (2010) Negative differential resistance and resistive switching behaviors in Cu_2S nanowire devices. Appl Phys Lett 96(22):223103. https://doi.org/10.1063/1.3442919
84. Liu JW, Xu J, Liang HW, Wang K, Yu SH (2012) Macroscale ordered ultrathin telluride nanowire films, and tellurium/telluride hetero-nanowire films. Angew Chem Int Ed 51 (30):7420–7425. https://doi.org/10.1002/anie.201201608
85. Seo HJ, Jeong W, Lee S, Moon GD (2018) Ultrathin silver telluride nanowire films and gold nanosheet electrodes for a flexible resistive switching device. Nanoscale 10(12):5424–5430. https://doi.org/10.1039/C8NR01429A
86. Zhai T, Li L, Wang X, Fang X, Bando Y, Golberg D (2010) Recent developments in one-dimensional inorganic nanostructures for photodetectors. Adv Func Mater 20(24):4233–4248. https://doi.org/10.1002/adfm.201001259
87. Velusamy DB, Kim RH, Cha S, Huh J, Khazaeinezhad R, Kassani SH, Song G, Cho SM, Cho SH, Hwang I, Lee J, Oh K, Choi H, Park C (2015) Flexible transition metal dichalcogenide nanosheets for band-selective photodetection. Nat Commun 6:8063. https://doi.org/10.1038/ncomms9063
88. Fang X, Bando Y, Liao M, Gautam UK, Zhi C, Dierre B, Liu B, Zhai T, Sekiguchi T, Koide Y, Golberg D (2009) Single-crystalline ZnS nanobelts as ultraviolet-light sensors. Adv Mater 21(20):2034–2039. https://doi.org/10.1002/adma.200802441
89. Zhai T, Fang X, Liao M, Xu X, Li L, Liu B, Koide Y, Ma Y, Yao J, Bando Y, Golberg D (2010) Fabrication of high-quality In_2Se_3 nanowire arrays toward high-performance visible-light photodetectors. ACS Nano 4(3):1596–1602. https://doi.org/10.1021/nn9012466
90. Wang J-J, Cao F-F, Jiang L, Guo Y-G, Hu W-P, Wan L-J (2009) High performance photodetectors of individual InSe single crystalline nanowire. J Am Chem Soc 131 (43):15602–15603. https://doi.org/10.1021/ja9072386

91. Jie JS, Zhang WJ, Jiang Y, Meng XM, Li YQ, Lee ST (2006) Photoconductive characteristics of single-crystal CdS nanoribbons. Nano Lett 6(9):1887–1892. https://doi.org/10.1021/nl060867g
92. Zhou R, Chang H-C, Protasenko V, Kuno M, Singh AK, Jena D, Xing H (2007) CdSe nanowires with illumination-enhanced conductivity: induced dipoles, dielectrophoretic assembly, and field-sensitive emission. J Appl Phys 101(7):073704. https://doi.org/10.1063/1.2714670
93. Fang X, Xiong S, Zhai T, Bando Y, Liao M, Gautam UK, Koide Y, Zhang X, Qian Y, Golberg D (2009) High-performance blue/ultraviolet-light-sensitive ZnSe-nanobelt photodetectors. Adv Mater 21(48):5016–5021. https://doi.org/10.1002/adma.200902126
94. Zhai T, Ye M, Li L, Fang X, Liao M, Li Y, Koide Y, Bando Y, Golberg D (2010) Single-crystalline Sb_2Se_3 nanowires for high-performance field emitters and photodetectors. Adv Mater 22(40):4530–4533. https://doi.org/10.1002/adma.201002097
95. Konstantatos G, Levina L, Tang J, Sargent EH (2008) Sensitive solution-processed Bi_2S_3 nanocrystalline photodetectors. Nano Lett 8(11):4002–4006. https://doi.org/10.1021/nl802600z
96. Xiao G, Dong Q, Wang Y, Sui Y, Ning J, Liu Z, Tian W, Liu B, Zou G, Zou B (2012) One-step solution synthesis of bismuth sulfide (Bi_2S_3) with various hierarchical architectures and their photoresponse properties. RSC Adv 2(1):234–240. https://doi.org/10.1039/C1RA00289A
97. Wang D, Hao C, Zheng W, Peng Q, Wang T, Liao Z, Yu D, Li Y (2008) Ultralong single-crystalline Ag_2S nanowires: promising candidates for photoswitches and room-temperature oxygen sensors. Adv Mater 20(13):2628–2632. https://doi.org/10.1002/adma.200800138
98. Zhang E, Wang P, Li Z, Wang H, Song C, Huang C, Chen Z-G, Yang L, Zhang K, Lu S, Wang W, Liu S, Fang H, Zhou X, Yan H, Zou J, Wan X, Zhou P, Hu W, Xiu F (2016) Tunable ambipolar polarization-sensitive photodetectors based on high-anisotropy $ReSe_2$ nanosheets. ACS Nano 10(8):8067–8077. https://doi.org/10.1021/acsnano.6b04165
99. Lopez-Sanchez O, Lembke D, Kayci M, Radenovic A, Kis A (2013) Ultrasensitive photodetectors based on monolayer MoS_2. Nat Nanotechnol 8:497. https://doi.org/10.1038/nnano.2013.100
100. Perea-López N, Elías AL, Berkdemir A, Castro-Beltran A, Gutiérrez HR, Feng S, Lv R, Hayashi T, López-Urías F, Ghosh S, Muchharla B, Talapatra S, Terrones H, Terrones M (2013) Photosensor device based on few-layered WS2 films. Adv Func Mater 23(44):5511–5517. https://doi.org/10.1002/adfm.201300760
101. Zhang B, Hou W, Ye X, Fu S, Xie Y (2007) 1D Tellurium nanostructures: photothermally assisted morphology-controlled synthesis and applications in preparing functional nanoscale materials. Adv Func Mater 17(3):486–492. https://doi.org/10.1002/adfm.200600566
102. Wang J-J, Wang Y-Q, Cao F-F, Guo Y-G, Wan L-J (2010) Synthesis of monodispersed Wurtzite structure $CuInSe_2$ nanocrystals and their application in high-performance organic −inorganic hybrid photodetectors. J Am Chem Soc 132(35):12218–12221. https://doi.org/10.1021/ja1057955
103. Wang J-J, Hu J-S, Guo Y-G, Wan L-J (2012) Wurtzite $Cu_2ZnSnSe_4$ nanocrystals for high-performance organic–inorganic hybrid photodetectors. Npg Asia Mater 4:e2. https://doi.org/10.1038/am.2012.2
104. Xue DJ, Wang JJ, Wang YQ, Xin S, Guo YG, Wan LJ (2011) Facile synthesis of germanium nanocrystals and their application in organic-inorganic hybrid photodetectors. Adv Mater 23(32):3704–3707. https://doi.org/10.1002/adma.201101436
105. Wang X, Song W, Liu B, Chen G, Chen D, Zhou C, Shen G (2013) High-performance organic-inorganic hybrid photodetectors based on P3HT:CdSe nanowire heterojunctions on rigid and flexible substrates. Adv Func Mater 23(9):1202–1209. https://doi.org/10.1002/adfm.201201786
106. Shao M-H, Sasaki K, Adzic RR (2006) Pd−Fe nanoparticles as electrocatalysts for oxygen reduction. J Am Chem Soc 128(11):3526–3527. https://doi.org/10.1021/ja060167d

107. Nørskov JK, Rossmeisl J, Logadottir A, Lindqvist L, Kitchin JR, Bligaard T, Jónsson H (2004) Origin of the overpotential for oxygen reduction at a fuel-cell cathode. J Phys Chem B 108(46):17886–17892. https://doi.org/10.1021/jp047349j
108. Winter M, Brodd RJ (2004) What are batteries, fuel cells, and supercapacitors? Chem Rev 104(10):4245–4270. https://doi.org/10.1021/cr020730k
109. Gao MR, Jiang J, Yu SH (2012) Solution-based synthesis and design of late transition metal chalcogenide materials for oxygen reduction reaction (ORR). Small 8(1):13–27. https://doi.org/10.1002/smll.201101573
110. Zhao W, Zhang C, Geng F, Zhuo S, Zhang B (2014) Nanoporous hollow transition metal chalcogenide nanosheets synthesized via the anion-exchange reaction of metal hydroxides with chalcogenide ions. ACS Nano 8(10):10909–10919. https://doi.org/10.1021/nn504755x
111. Gao M-R, Liu S, Jiang J, Cui C-H, Yao W-T, Yu S-H (2010) In situ controllable synthesis of magnetite nanocrystals/$CoSe_2$ hybrid nanobelts and their enhanced catalytic performance. J Mater Chem 20(42):9355–9361. https://doi.org/10.1039/C0JM01547D
112. Checkelsky JG, Hor YS, Cava RJ, Ong NP (2011) Bulk band gap and surface state conduction observed in voltage-tuned crystals of the topological insulator Bi_2Se_3. Phys Rev Lett 106(19):196801. https://doi.org/10.1103/PhysRevLett.106.196801
113. Kong D, Chen Y, Cha JJ, Zhang Q, Analytis JG, Lai K, Liu Z, Hong SS, Koski KJ, Mo S-K, Hussain Z, Fisher IR, Shen Z-X, Cui Y (2011) Ambipolar field effect in the ternary topological insulator $(BixSb_{1-x})_2Te_3$ by composition tuning. Nat Nanotechnol 6:705. https://doi.org/10.1038/nnano.2011.172
114. Fu L, Kane CL, Mele EJ (2007) Topological insulators in three dimensions. Phys Rev Lett 98(10):106803. https://doi.org/10.1103/PhysRevLett.98.106803
115. Garate I, Franz M (2010) Inverse spin-galvanic effect in the interface between a topological insulator and a ferromagnet. Phys Rev Lett 104(14):146802. https://doi.org/10.1103/PhysRevLett.104.146802
116. Fu L, Kane CL (2008) Superconducting proximity effect and majorana fermions at the surface of a topological insulator. Phys Rev Lett 100(9):096407. https://doi.org/10.1103/PhysRevLett.100.096407
117. Chen YL, Analytis JG, Chu J-H, Liu ZK, Mo S-K, Qi XL, Zhang HJ, Lu DH, Dai X, Fang Z, Zhang SC, Fisher IR, Hussain Z, Shen Z-X (2009) Experimental realization of a three-dimensional topological insulator, Bi_2Te_3. Science 325(5937):178–181. https://doi.org/10.1126/science.1173034
118. Zhang H, Liu C-X, Qi X-L, Dai X, Fang Z, Zhang S-C (2009) Topological insulators in Bi_2Se_3, Bi_2Te_3 and Sb_2Te_3 with a single Dirac cone on the surface. Nat Phys 5:438. https://doi.org/10.1038/nphys1270
119. Xia Y, Qian D, Hsieh D, Wray L, Pal A, Lin H, Bansil A, Grauer D, Hor YS, Cava RJ, Hasan MZ (2009) Observation of a large-gap topological-insulator class with a single Dirac cone on the surface. Nat Phys 5:398. https://doi.org/10.1038/nphys1274
120. Davies J (1998) The physics of low-dimensional semiconductors: an introduction. Cambridge University Press, New York
121. Kong D, Randel JC, Peng H, Cha JJ, Meister S, Lai K, Chen Y, Shen Z-X, Manoharan HC, Cui Y (2010) Topological insulator nanowires and nanoribbons. Nano Lett 10(1):329–333. https://doi.org/10.1021/nl903663a
122. Zhang W, Yu R, Feng W, Yao Y, Weng H, Dai X, Fang Z (2011) Topological aspect and quantum magnetoresistance of β-Ag_2Te. Phys Rev Lett 106(15):156808. https://doi.org/10.1103/PhysRevLett.106.156808
123. Hutter E, Fendler JH (2004) Exploitation of localized surface plasmon resonance. Adv Mater 16(19):1685–1706. https://doi.org/10.1002/adma.200400271
124. Homola J (2003) Present and future of surface plasmon resonance biosensors. Anal Bioanal Chem 377(3):528–539. https://doi.org/10.1007/s00216-003-2101-0
125. Homola J (2008) Surface plasmon resonance sensors for detection of chemical and biological species. Chem Rev 108(2):462–493. https://doi.org/10.1021/cr068107d

126. Nelson BP, Grimsrud TE, Liles MR, Goodman RM, Corn RM (2001) Surface plasmon resonance imaging measurements of DNA and RNA hybridization adsorption onto DNA microarrays. Anal Chem 73(1):1–7. https://doi.org/10.1021/ac0010431
127. Brongersma ML, Hartman JW, Atwater HA (2000) Electromagnetic energy transfer and switching in nanoparticle chain arrays below the diffraction limit. Phys Rev B 62(24): R16356–R16359. https://doi.org/10.1103/PhysRevB.62.R16356
128. Ozbay E (2006) Plasmonics: merging photonics and electronics at nanoscale dimensions. Science 311(5758):189–193. https://doi.org/10.1126/science.1114849
129. Mattox TM, Ye X, Manthiram K, Schuck PJ, Alivisatos AP, Urban JJ (2015) Chemical control of plasmons in metal chalcogenide and metal oxide nanostructures. Adv Mater 27 (38):5830–5837. https://doi.org/10.1002/adma.201502218
130. Hsu S-W, On K, Tao AR (2011) Localized surface plasmon resonances of anisotropic semiconductor nanocrystals. J Am Chem Soc 133(47):19072–19075. https://doi.org/10.1021/ja2089876
131. Dorfs D, Härtling T, Miszta K, Bigall NC, Kim MR, Genovese A, Falqui A, Povia M, Manna L (2011) Reversible tunability of the near-infrared valence band plasmon resonance in $Cu_{2-x}Se$ Nanocrystals. J Am Chem Soc 133(29):11175–11180. https://doi.org/10.1021/ja2016284
132. Llordés A, Garcia G, Gazquez J, Milliron DJ (2013) Tunable near-infrared and visible-light transmittance in nanocrystal-in-glass composites. Nature 500:323. https://doi.org/10.1038/nature12398
133. Mattox TM, Bergerud A, Agrawal A, Milliron DJ (2014) Influence of shape on the surface plasmon resonance of tungsten bronze nanocrystals. Chem Mater 26(5):1779–1784. https://doi.org/10.1021/cm4030638
134. Niezgoda JS, Harrison MA, McBride JR, Rosenthal SJ (2012) Novel synthesis of chalcopyrite $CuxInyS_2$ quantum dots with tunable localized surface plasmon resonances. Chem Mater 24(16):3294–3298. https://doi.org/10.1021/cm3021462
135. Hsu S-W, Bryks W, Tao AR (2012) Effects of carrier density and shape on the localized surface plasmon resonances of $Cu_{2-x}S$ nanodisks. Chem Mater 24(19):3765–3771. https://doi.org/10.1021/cm302363x
136. Zhao Y, Pan H, Lou Y, Qiu X, Zhu J, Burda C (2009) Plasmonic $Cu_{2-x}S$ nanocrystals: optical and structural properties of copper-deficient Copper(I) Sulfides. J Am Chem Soc 131 (12):4253–4261. https://doi.org/10.1021/ja805655b
137. Kriegel I, Jiang C, Rodríguez-Fernández J, Schaller RD, Talapin DV, da Como E, Feldmann J (2012) Tuning the excitonic and plasmonic properties of copper chalcogenide nanocrystals. J Am Chem Soc 134(3):1583–1590. https://doi.org/10.1021/ja207798q
138. Nath M, Kar S, Raychaudhuri AK, Rao CNR (2003) Superconducting $NbSe_2$ nanostructures. Chem Phys Lett 368(5):690–695. https://doi.org/10.1016/S0009-2614(02)01930-9
139. Dunnill CW, Edwards HK, Brown PD, Gregory DH (2006) Single-step synthesis and surface-assisted growth of superconducting TaS_2 nanowires. Angew Chem Int Ed 45 (42):7060–7063. https://doi.org/10.1002/anie.200602614
140. Gresty NC, Takabayashi Y, Ganin AY, McDonald MT, Claridge JB, Giap D, Mizuguchi Y, Takano Y, Kagayama T, Ohishi Y, Takata M, Rosseinsky MJ, Margadonna S, Prassides K (2009) Structural phase transitions and superconductivity in Fe1+δSe0.57Te0.43 at ambient and elevated pressures. J Am Chem Soc 131(46):16944–16952. https://doi.org/10.1021/ja907345x
141. Mizuguchi Y, Tomioka F, Tsuda S, Yamaguchi T, Takano Y (2009) Superconductivity in S-substituted FeTe. Appl Phys Lett 94(1):012503. https://doi.org/10.1063/1.3058720
142. Hsu F-C, Luo J-Y, Yeh K-W, Chen T-K, Huang T-W, Wu PM, Lee Y-C, Huang Y-L, Chu Y-Y, Yan D-C, Wu M-K (2008) Superconductivity in the PbO-type structure α-FeSe. Proc Natl Acad Sci 105(38):14262–14264. https://doi.org/10.1073/pnas.0807325105
143. Sales BC, Sefat AS, McGuire MA, Jin RY, Mandrus D, Mozharivskyj Y (2009) Bulk superconductivity at 14 K in single crystals of $Fe_{1+y}TexSe_{1-x}$. Phys Rev B 79(9):094521. https://doi.org/10.1103/PhysRevB.79.094521

144. Medvedev S, McQueen TM, Troyan IA, Palasyuk T, Eremets MI, Cava RJ, Naghavi S, Casper F, Ksenofontov V, Wortmann G, Felser C (2009) Electronic and magnetic phase diagram of β-Fe1.01Se with superconductivity at 36.7 K under pressure. Nat Mater 8:630. https://doi.org/10.1038/nmat2491
145. Mizuguchi Y, Tomioka F, Tsuda S, Yamaguchi T, Takano Y (2009) Substitution effects on FeSe superconductor. J Phys Soc Jpn 78(7):074712. https://doi.org/10.1143/JPSJ.78.074712
146. Guo J, Jin S, Wang G, Wang S, Zhu K, Zhou T, He M, Chen X (2010) Superconductivity in the iron selenide $KxFe_2Se_2$ ($0 \leq x \leq 1.0$). Phys Rev B 82(18):180520. https://doi.org/10.1103/physrevb.82.180520
147. Malavasi L, Margadonna S (2012) Structure-properties correlations in Fe chalcogenide superconductors. Chem Soc Rev 41(10):3897–3911. https://doi.org/10.1039/C2CS35021A

Chapter 4
Conclusion and Perspectives

Abstract This brief presented an overview of syntheses and applications of anisotropic MC nanomaterials. Tuning the morphologies of MC nanomaterials is categorized into four different routes: intrinsic growth, shape-guiding agent growth, oriented growth, and chemical transformation. Assembly of the pre-formed anisotropic MC nanomaterials widens their application by building up thin layer or superstructures. In addition, compositional variations in MC nanomaterials has been leading to a variety of applications, which can be further developed into new and unexplored morphologies in the near future.

This brief covered synthetic and assembly strategies of anisotropic metal chalcogenide nanocrystals and their applications. The synthesis part reviewed four different approaches to synthesize anisotropic MC nanocrystals; intrinsic growth into 1D or 2D nanostructure, crystal growth with shape-guiding agents, oriented attachment, and chemical transformation. In addition, atomically-thin films of MCs were also introduced because they are on the center of the MC research due to its quantum properties with their availability of large area fabrication. This application part provides representative examples of anisotropic MC nanomaterials that are expected to be meaningful, academically and/or industrially, in the near future. The applications include energy storage and conversion, memory device, photodetectors, electrocatalyst, topological insulator, localized surface plasmon resonance, and superconductor. Nevertheless, a number of challenges should be addressed in the solution-based synthesis of the anisotropic MC nanocrystals to realize practical applications in industry. There are critical issues commonly in the way of targeted and well-designed MC nanostructure fabrication; (1) precise understanding on the surface energy decrease by surfactants, (2) adopting chemical transformations in the nucleation and growth of MC nanocrystals, (3) removal of organic surfactant, (4) kinetic study on the chemical transformation, (5) scale-up of production to meet the industrial need, and (6) eco-friendly synthesis.

In the synthesis utilizing organic surfactants, the decrease in surface energy depends on the number of binding and the binding energy between the surfactant

G. D. Moon, *Anisotropic Metal Chalcogenide Nanomaterials*,
SpringerBriefs in Materials, https://doi.org/10.1007/978-3-030-03943-1_4

molecules and the inorganic surfaces. Binding exclusively to a specific surface can facilitate the oriented attachment to form nanowires or nanoplates, but binding to all the surfaces helps self-assembly of the nanocrystals into super-structured materials. In the layer-structured materials, the number of crystal surfaces exposed to the solution is limited (typically 2 surfaces) and the surfactant molecules adsorb on the top and bottom surfaces without binding to the side surface. The small flat nanocrystals go through the oriented attachment side-by-side and form larger nanoplates or nanosheets, hence oriented attachment in 2D structures is natural as seen in many recent publications [1–3]. On the contrary, the oriented attachment for generating 1D nanomaterials requests delicate control of binding of surfactant molecules. The small building block nanocrystals to be used for this oriented attachment may be isotropic or shows small deviation from an isotropic shape. To succeed in exclusive binding of surfactant molecules to a specific surface of the building block nanocrystals, we need precise quantitative information on the binding energy and the resulting surface energy decrease. Unfortunately, the information is not easy to obtain because the binding is affected by the reaction conditions in complex and mixed ways with surfactant molecules, precursors, solvent, temperature, and pH, etc. Additional issue regarding to the surfactant is the assembly of nanocrystals that are driven by interaction between the surfactant molecules. This surfactant interaction may induce self-assembly of the building blocks to form super-structured materials or cause the formation of templates such as micelles or vesicles consisting of the surfactant molecules and the MC elements. Currently, it is not clear which molecules can exist as the organometal complex and form such templates during the synthesis. Several anisotropic MCs have been prepared by the template-based synthesis, but the physics on the formation of such templates needs thorough investigation. The physics is more complicated than the micelle formation of pure organic molecules because the interaction between the inorganic elements incorporated in the organic molecules should be taken into consideration in the assembly of the complex molecules.

Another challenge in the synthesis of anisotropic MC nanocrystals is the lack of knowledge on how the chemical composition of the nanocrystals is determined. The theory of conventional nucleation and growth in a solution phase has been established under the assumption of stoichiometric supply of elemental sources. Concentration of a source element is determined by the reduction rate of the precursor. Finding a synthetic condition at which the reduction rates of the source molecules are the same is critical in the nanocrystal synthesis. It is not certain whether the elemental composition of a product nanocrystal is identical to the elemental concentration of the sources in the nucleation and growth steps. Because the equilibrium constants of the precursors are different and vary during the synthesis, it is almost impossible to maintain the concentration of the source elements stoichiometric. This hypothesis leads to the existence of chemical transformation taking place during the nanocrystal synthesis. In the MC nanocrystals, the in situ chemical transformation is the alloy formation [4]. Taking an example with CdSe quantum dots, the nuclei might be Cd-rich or Se-rich or even pure Cd or Se, and then the other element forms alloy nanocrystals with the stoichiometric composition of 1:1. The alloy formation is spontaneous because the alloy composition is thermodynamically more stable. As long as initial nuclei do not grow into anisotropic structure, the overall shape of the final product will be 0D and the precise facets will be determined by surface energy of the nanocrystal. Hence, the shape of 0D nanocrystals is not very sensitive to reaction temperature. In the anisotropic MC nanocrystals, however, the fast anisotropic growth of one component can result in completely different shape of final product although the chemical composition of the nanocrystals is identical. For example, The shape of Bi_2Te_3 is strongly dependent on the reaction temperature, typically varying from 2D nanoplates at high temperatures (~200 °C) to nanowires with lots of nanoplates growing in the radial direction at lower temperatures (~150 °C), or to nanowires with multiple grains at much lower temperatures (~100 °C) [5]. At the high temperatures, the elemental fraction of the reduced Bi and Te is considered similar to 2:3, hence the stable stoichiometric 2D nanoplates could be nucleated and grown. At the low temperatures, Te grows first into the nanorods or nanowires because of its intrinsic preference to 1D structure, and then they were transformed into Bi_2Te_3. Shape of the final product should depend on the length of pure Te nanowires before Bi started to form alloy. This result is an example showing the importance of understanding the reaction kinetics in the shape control of anisotropic nanocrystals. Unfortunately, the kinetics has not been studied thoroughly. For the ternary or quaternary nanocrystals, monitoring the reaction kinetics is a very difficult task at the current stage. Data base on the reduction rate of the precursors in a variety of solution conditions should be accumulated by both experimental and theoretical studies.

Surfactants are inevitable in preparation of stable colloidal suspension and in the shape control of MC nanocrystals. However, the organic surfactants are problematic in practical uses. Many applications involve sintering of the MC nanocrystal films which aims at preparation of nanograined composites (thermoelectric devices) or simply at solution processing for cost-effective device fabrication (CI(G)S solar cell). The insulating organic layer plays as a barrier for charge injection and diminishes the expected properties from the pure MCs. The surfactant molecules change the surface electronic state of the individual MC nanocrystals, hence studies on the topological insulator and the high-mobility semiconductor have been conducted with the MC films obtained by the dry-processes. Nanocrystal thick films often exhibit weak mechanical strength caused by the low material density and micropores generated during the sintering the nanocrystal films. The problem regarding the surfactant can be reduced by the synthesis of surfactant-free nanocrystals, ligand exchange by volatile organic molecules, or the use of inorganic surfactants. The surfactant-free synthesis is promising for the 2D MC nanocrystals because their basal surfaces are covered with chalcogen atoms. The slightly negative charge of the chalcogen atoms can provide colloidal stability of the nanocrystals. Although the stability without surfactant is limited, the nanoplates can be redispersed in the solution by simple ultrasonication and formulated into nanocrystal inks. If the application is to produce thick films of MC crystals through thermal treatment, slurry-like ink is useful for the coating process. When the long-term stability is critical, the organic surfactant can be replaced by volatile molecules before being used. Employing inorganic ligands looks promising because such surfactants do not form the insulating barrier layer on the nanocrystal surfaces. The approach allows low-temperature thermal annealing which facilitates fabrication of flexible devices on polymer substrates. Furthermore, the inorganic surfactants are expected to maintain the high charge mobility and topological insulating nature of the 2D MC nanocrystals. Recently, molecular metal chalcogenide complexes (MCCs) based on Sn or Cu metal element have been successfully employed as the inorganic surfactants [6, 7]. The other MCs should be developed for diverse MC compositions.

Precise shape control and heterostructure design are the main advantages of the solution-based synthesis. When it comes to the amount of production, mechanical milling of bulk MCs into a fine powder is more profitable in the cost. Because most applications of the anisotropic MC nanocrystals use thick films sintered at high temperatures, the gram scale production is not sufficient. For example, one pellet specimen for thermoelectric test requests about 0.5 g of nanocrystals. The CIGS solar module or a set of LIB coin cell electrodes also need similar amount for each test, which should be scaled up to kg production per batch for industrial use. Obstacles to achieve the massive production are the relatively low solubility of the MC precursors, pressure build-up by the gaseous by-products during the synthesis, and inhomogeneous mixing in large scale reactors which is problematic in fast reaction systems. Continuous-flow reaction systems can be an alternative way to the batch systems. Fast mixing, easy temperature control, and small amount consumption of reagents are advantageous over the batch-type reaction. Recently, Seeberger and coworkers demonstrated a microfluidic system to produce isotropic quantum dots [8]. Jeon and coworkers reported the synthesis of ZnSe@ZnS core-shell quantum dots in a microfluidic reaction system [9]. Production of anisotropic nanomaterials has not been reported in the continuous reaction systems. Scale-up of the continuous-flow reaction and the parallel integration of the set-up may facilitate massive production to meet the industrial need.

Nanocrystal inks are gaining tremendous attention to achieve cost-effective coating process. Eco-friendly synthesis becomes important. Currently, MC nanocrystals are synthesized mainly by the solvothermal approach in which the reaction is conducted at a high temperature in organic solvents. The use of solvents with a high boiling temperature in the approach makes it difficult to recycle the solvents. Synthesis in water or volatile alcohols is desirable in terms of the environment friendliness. The reaction yield of 100% is also an important issue in the inorganic synthesis in order to prevent contamination by toxic elements. Especially, the chemical transformation approach uses excess amount of a source material for exchange reaction when the transformation is not favored in energetics [10]. Quite recently, ultrathin 2D layered MC nanomaterials have attracted a lot of attention in a way that expands from binary MC into ternary (Ta_2NiS_5, Cu_2WS_4), quaternary, alloyed ($TaS_{2x}S_{2(1-x)}$, $Ti_xTa_{1-x}S_2$), heteroatom-doped, and heterostructures MC nanostructures (ZrS_2–ReS_2). These ultrathin 2D MC nanomaterials have potential applications including sodium ion battery, electrocatalytic CO_2 reduction, and water remediation in addition to electronics/optoelectronics, catalysis, sensors, energy storage/conversion. All in all, the future direction of MC lies in the development of new potential applications in parallel with new discovery of unexplored properties.

References

1. Min Y, Moon GD, Kim BS, Lim B, Kim J-S, Kang CY, Jeong U (2012) Quick, controlled synthesis of ultrathin Bi_2Se_3 nanodiscs and nanosheets. J Am Chem Soc 134(6):2872–2875. https://doi.org/10.1021/ja209991z
2. Schliehe C, Juarez BH, Pelletier M, Jander S, Greshnykh D, Nagel M, Meyer A, Foerster S, Kornowski A, Klinke C, Weller H (2010) Ultrathin PbS sheets by two-dimensional oriented attachment. Science 329(5991):550–553. https://doi.org/10.1126/science.1188035
3. Li L, Chen Z, Hu Y, Wang X, Zhang T, Chen W, Wang Q (2013) Single-layer single-crystalline SnSe nanosheets. J Am Chem Soc 135(4):1213–1216. https://doi.org/10.1021/ja3108017
4. Moon GD, Ko S, Min Y, Zeng J, Xia Y, Jeong U (2011) Chemical transformations of nanostructured materials. Nano Today 6(2):186–203. https://doi.org/10.1016/j.nantod.2011.02.006
5. Zhang Y, Hu LP, Zhu TJ, Xie J, Zhao XB (2013) High yield Bi_2Te_3 single crystal nanosheets with uniform morphology via a solvothermal synthesis. Cryst Growth Des 13(2):645–651. https://doi.org/10.1021/cg3013156
6. Kovalenko MV, Scheele M, Talapin DV (2009) Colloidal nanocrystals with molecular metal chalcogenide surface ligands. Science 324(5933):1417–1420. https://doi.org/10.1126/science.1170524
7. Liu W, Lee J-S, Talapin DV (2013) III–V nanocrystals capped with molecular metal chalcogenide ligands: high electron mobility and ambipolar photoresponse. J Am Chem Soc 135(4):1349–1357. https://doi.org/10.1021/ja308200f
8. Laurino P, Kikkeri R, Seeberger PH (2011) Continuous-flow reactor–based synthesis of carbohydrate and dihydrolipoic acid–capped quantum dots. Nat Protoc 6:1209. https://doi.org/10.1038/nprot.2011.357
9. Kwon BH, Lee KG, Park TJ, Kim H, Lee TJ, Lee SJ, Jeon DY (2012) Continuous in situ synthesis of ZnSe/ZnS core/shell quantum dots in a microfluidic reaction system and its application for light-emitting diodes. Small 8(21):3257–3262. https://doi.org/10.1002/smll.201200773
10. Moon GD, Ko S, Xia Y, Jeong U (2010) Chemical transformations in ultrathin chalcogenide nanowires. ACS Nano 4(4):2307–2319. https://doi.org/10.1021/nn9018575

Printed in the United States
By Bookmasters